DO-IT-YOURSELF

FLOORING

BY THE EDITORS OF SUNSET BOOKS AND SUNSET MAGAZINE

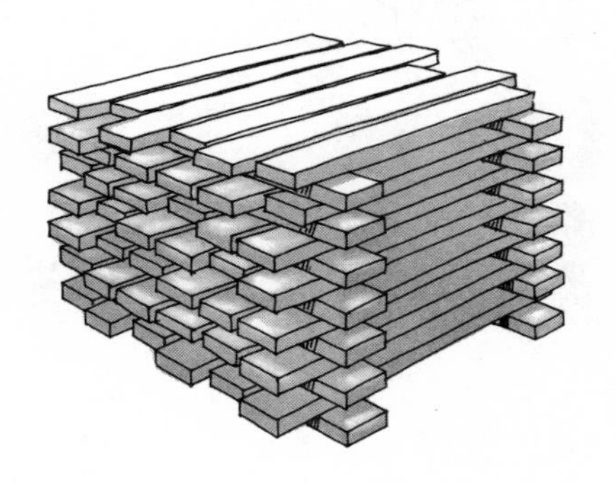

LANE PUBLISHING CO., MENLO PARK, CALIFORNIA

A team effort
went into the preparation of this book. For special help in checking and advising on technical information, we offer our thanks to the following people, companies, and organizations: Circus Floors; Edward Clifford; Color Tile; Color Tile Supermart, Inc.; Morris Neil Finisy; Scott Lundberg; Dwight Marsh; National Oak Flooring Manufacturer's Association; Rex Floorcoverings; Stuart Floor Co.; Tile Council of America; and Donald W. Vandervort. A special note of thanks to Hilary Hannon, who helped research the color section.

Photographers

Franciscan Tile: 23 right. **Jack McDowell:** 10 top & bottom right, 12 right, 22 left, 23 left, 26 top & bottom left, 28 top, 30 right, 35, 38 bottom left & right, 40. **Steve W. Marley:** 10 bottom left, 11, 14 left, 15 right, 19, 25, 27, 30 left, 32, 37 left, 39. **Jim Peck:** 28 bottom. **Rob Super:** 9 top, 13 right, 14 right, 18 right, 22 right, 24 top, 26 bottom right. **Tom Wyatt:** 9 bottom, 12 left, 13 left, 15 left, 16, 17, 18 left, 20, 21, 24 bottom left & right, 29, 31, 33, 34, 36, 37 right, 38 top.

Editor, Sunset Books: David E. Clark

First printing May 1982

 Library of Congress No. 81-82872. ISBN 0-376-01141-6. Lithographed in the United States.

Supervising Editor:
Sherry Gellner

Research & Text:
Bill Tanler

Staff Editors: **Barbara J. Braasch**
Don Rutherford

Photo Editors: **Scott Fitzgerrell**
JoAnn Masaoka Lewis

Design: **Joe di Chiarro**

Illustrations: **Rik Olson**

Cover: Photograph by **Steve W. Marley.** Cover design by **Zan Fox**

CONTENTS

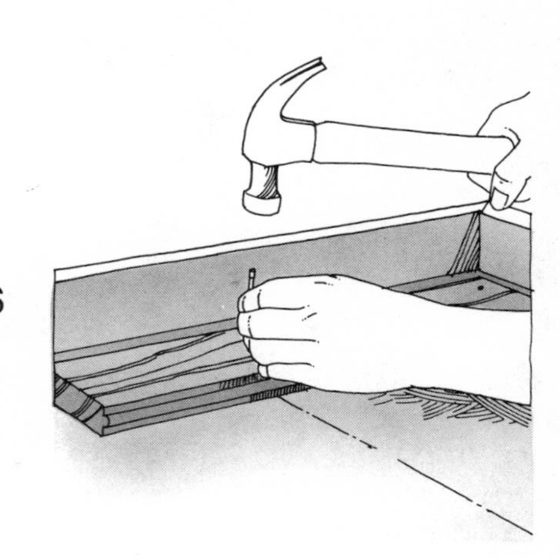

YOUR CHOICES IN FLOORING MATERIALS

A myriad of types and styles to choose from—each with its own character and quality

- guidelines for making your selection
- types of flooring materials

Wood, resilient flooring, ceramic tile, masonry, and carpeting—all are available today in a wide variety of styles and qualities, and all are good flooring choices. New flooring materials, effective adhesives, and durable sealers and finishes make it possible to select any type of flooring for any climate and geographical location.

To become familiar with the different types of flooring available and choose one that both complements the interior design of your home and meets your needs, read the descriptions of the flooring choices beginning on the facing page. Then explore the color gallery that begins on page 9; you'll discover innovative and imaginative uses of floor coverings that can help spark your imagination.

If you're planning to do the work yourself, you'll want to know exactly what's involved in laying the particular kind of flooring you've chosen. Complete instructions for installing various flooring materials are in the section beginning on page 41. Following the installation techniques are detailed directions for repairing and refinishing old floors.

It's a good idea to take the time to visit flooring dealers, home improvement centers, and flooring material suppliers to see the materials available in your area. Most dealers are happy to let you borrow samples so you can see how they look in your own home.

GUIDELINES FOR MAKING YOUR SELECTION

Once you're familiar with the range of possibilities for flooring, you'll be able to narrow down your choices based on your needs and your own taste. Here are some important guidelines to keep in mind as you're making your final decision.

• **Esthetics.** The flooring material's design, texture, even the way it feels underfoot can set the mood of a room, even of an entire house. An area carpeted in a low, thick plush becomes a warm and inviting gathering place. A room laid with bold masonry blocks can appear to bring the outdoors into the house.

Make sure the flooring you choose both communicates the feeling of the room and complements its decor. Since good quality flooring will last for many years—often as long as the house itself—you'll want to avoid designs, colors, or patterns that you may tire of quickly.

• **Wear.** Determine the kind and amount of traffic the flooring is expected to bear. For high-traffic areas, such as hallways, entryways, kitchens, and bathrooms, select the most durable materials you can find. Areas that will receive less wear can be covered with less rugged grades of material.

• **Cost.** All the standard flooring materials come in various grades, with the cost directly related to the quality. Generally, the best quality materials provide the longest and most satisfactory service, and are the most economical in the long run. A good rule of thumb is to install the best quality you can afford.

• **Comfort.** Some flooring materials will be softer underfoot than others. If you object to a cold, hard surface, you may not want to use ceramic tile or masonry materials. Wood or resilient flooring provides a less firm surface. Softest of all is carpeting.

• **Noise.** Soft flooring materials, such as vinyl, rubber, and carpeting, deaden sound. Wood, ceramic tile, masonry, and other hard surfaces tend to reflect sound, rather than absorb it.

• **Safety.** Avoid slippery finishes whenever possible. In kitchens and bathrooms, don't use flooring that becomes slick when damp. Loose rugs laid on hardwood floors should have nonskid backings or pads.

• **Installation.** Most types of flooring can be installed by a competent do-it-yourselfer. The installation section beginning on page 41 explains the techniques used for each kind of flooring. If you want to do the work yourself, be sure to check the instructions first to see if the flooring that interests you is within the range of your abilities.

• **Maintenance.** New floor materials, protective finishes, and cleaning techniques make maintenance less of a consideration now than in the past.

Still, flooring that's heavily traveled will get dirty and eventually may scratch or scuff.

Care and maintenance instructions for the various flooring materials are in each installation section; take time to read them now so you'll know how much work you'll need to do to keep your new or refinished floor in mint condition.

• **Subfloors.** Don't make any final flooring decision until you know the kind and condition of the subfloor your new flooring will cover. With proper preparation, a concrete subfloor—because it's rigid—can serve as a base for almost any type of flooring. Other subfloors are more flexible and not suitable for rigid materials, such as masonry and ceramic tile.

The installation section of this book will tell you not only the type of subfloor required for each flooring material, but also the preparations and repairs you may need to make on your subfloor before you can begin laying your new flooring.

TYPES OF FLOORING MATERIALS

Understanding the basic characteristics of the various materials and how they can be used will help you make your decision. On the following pages we present brief descriptions of five basic flooring choices so you can compare the advantages and disadvantages of each.

Wood

Despite the development of far less expensive synthetics, wood flooring remains as popular today as it was in the past.

Floors made from wood are usually called hardwood floors, even though they're sometimes made from softwoods such as fir, pine, hemlock, and redwood. Commonly used hardwoods include oak, maple, birch, and beech.

Wood floors that are properly sealed resist stains, scuffs, and scratches. When the floor is worn, it can be refinished to look like new.

High-grade hardwood is expensive, but the cost is somewhat offset by its durability. Wood floors may shrink in heat or swell in dampness. They require a very carefully prepared subfloor and a moisture-free environment.

Almost all wood flooring can be classified into three basic types—strip, plank, and wood block. Each has its own character and special uses.

Strip flooring. The most common wood flooring found in homes, strip flooring is composed of narrow tongue-and-groove boards laid in random lengths; the widths don't vary.

Strip flooring is available with tongue-and-groove ends or butt ends.

Plank flooring. Wide hardwood and softwood boards were readily available to our colonial ancestors. Maple planks more than a foot wide and secured to the subfloor with pegs can still be found in New England homes built in the 18th and 19th centuries.

Today, plank flooring—boards produced in random widths and random lengths—is still available. Unlike the planks of old, boards are now commonly milled with tongue-and-groove edges; the pegs you sometimes see in modern plank flooring are usually decorative touches.

Wood block flooring. Produced in dozens of patterns, textures, and thicknesses, block flooring is made from solid pieces of wood, laminates, or individual pieces of wood held together by a backing; floors made from small pieces are often called parquet or mosaic. Though usually manufactured in squares, blocks can also be rectangular.

Block flooring is easy to install, yet has the look of a custom-designed floor.

Resilient

The development of resins and synthetics has created a family of floor coverings described as resilient flooring. One of the earliest examples is linoleum, made from ground wood, cork, linseed oil, and resins. But the use of linoleum has been largely superseded by newer polyurethane, vinyl, and rubber materials that are flexible, moisture and stain-resistant, easy to install, and simple to maintain.

One of the advantages of resilient flooring is the seemingly endless variety of colors, textures, patterns, and styles you can choose from. Cost will vary widely, depending on the material.

Keep in mind, though, that resilient flooring is relatively soft, making it vulnerable to dents and tears caused by sharp objects, heavy furniture, and large appliances. But often such damage can be repaired.

Available either in sheets up to 12 feet wide or in 9-inch (asphalt only) or 12-inch-square tiles, resilient flooring is easy to handle. Installation requires patience and care, but the job is within the abilities of most homeowners.

Sheet flooring can cover an average-size room without seams, making it a practical flooring choice for kitchens, bathrooms, laundry rooms, and other areas where water spillage may be a problem.

Resilient tiles are easy to install because of their uniform size, but care must be taken to lay them squarely to avoid gaps and irregularities between tiles. Most resilient tiles are laid in adhesive, but some are available with self-sticking backing.

Sheet flooring. The versatility, attractiveness, and wide availability of sheet flooring make it a very popular floor covering, particularly in kitchens. Sheet flooring can be made from polyurethane, vinyl, or linoleum.

Polyurethane sheet flooring has a tough surface that keeps its luster for years with virtually no maintenance other than regular cleaning.

Vinyl sheet flooring has a top layer of vinyl bonded to a backing, usually felt or asbestos. When a resilient layer is pressed between the top layer and the backing, it produces cushioned vinyl—comfortable to walk on, durable, and quiet.

Linoleum now is less porous and longer wearing than in the past, but because it's often as expensive as sheet vinyl, it's no longer widely used or readily available.

Tile flooring. Tiles come in a wide range of colors, textures, patterns, and materials, including solid vinyl, vinyl-asbestos, rubber, and asbestos.

Vinyl tiles are manufactured with their patterns running the full thickness of the tiles. Their surface characteristics are the same as those for sheet vinyl.

Vinyl-asbestos tiles are made from asbestos filler bonded between two layers of vinyl. Though resistant to indentation, vinyl-asbestos tile is generally more susceptible to abrasion and damage from household chemicals than regular vinyl tile.

Rubber tile, made from synthetic rubber or a combination of synthetic and real rubber, was widely used for industrial flooring before it was considered for residential use. Long-wearing, quiet, waterproof, and resilient, rubber tiles are available in solid colors with ribbed or studded surfaces.

Asphalt tiles are not resistant to the penetration of grease and oil. For this reason, they're not recommended for use in kitchens.

Ceramic tile

Today, ceramic tile is a practical flooring for almost any room of the house. Made from hard-fired slabs of clay, tile is available in limitless patterns, often deep and brilliant colors, a variety of shapes, and several finishes.

Floor tiles come glazed and unglazed. The finish on glazed tiles may be glossy, satinlike-matte, semimatte, or dull. Unglazed tiles have no finish; the color, usually earth tones, is in the body of the tile. You can seal unglazed tiles after they're installed for easy maintenance.

The durability, easy care, and attractiveness of tile flooring are definite advantages, but the surface is very hard and reflects noise. In addition, tile is rigid and inflexible, it must be laid either on a concrete slab or on a structurally strong and sound wood subfloor.

Floor tiles are usually classified as quarry tile, pavers, patio tile, or glazed tile.

Quarry tiles. Though available with colorful glazed surfaces, unglazed red-clay quarry tiles are the most common. Quarry tiles are rough and water-resistant. They're laid in adhesives, and the joints are grouted.

Pavers. Like quarry tiles, pavers are normally unglazed and come in earthtone and other

(Continued on page 8)

SPECIAL EFFECTS—LET YOUR IMAGINATION GO

Using basic flooring materials and a little imagination, you can create special flooring effects that are interesting and unusual, and may have practical merits, as well.

The development of durable and tough paints, stains, and other floor finishes makes painted or stenciled wood floors practical. Squares of resilient tile can be arranged to create a giant checkerboard on a game room floor; smaller ceramic tiles can produce strikingly colorful patterns in custom designs. Even industrial floor coverings, once considered suitable only for factories or public areas like air terminals, are finding acceptance in homes, where their toughness is considered an advantage.

Here we introduce you to just a few ways in which you can adapt standard materials to special and unique applications. Use these ideas—as well as the striking special effects pictured on pages 36–39—as a starting point for your own explorations.

The many faces of wood

By their very character, wood floors provide a natural design in textures and patterns. Both strip and plank flooring have their own intrinsic design; often, this design can be enhanced by the way in which the strips or planks are laid (see pages 9–11).

You can create quite a different look by making a new wood block floor from scrap lumber cut in end-grain blocks. Special effects in existing wood floors can be produced by use of durable paints, colorful stains, special finishes, or stencil patterns.

You'll find information on the various kinds of wood floors, as well as installation instructions, beginning on page 42.

Resilient tile effects–from sophisticated to fun-loving

The many colors of resilient tile and ease of cutting it make it possible to create many special patterns for specific needs. The design can be as simple as a narrow border around the room in an accent color, or as playful as a ticktacktoe layout.

Innovative use of resilient tile involves little more than careful planning. Make a scale drawing on graph paper and sketch in the design you've chosen. Select graph paper that allows you to use one square on the paper for each piece of tile. Using the appropriate colors, fill in the squares so you can determine how much tile to order in each color.

Transfer the pattern to the floor so you can set tiles of the right color in their proper position. For complete installation instructions, see page 63.

Artistry with ceramic tile

From ancient times, skilled tile setters have created works of art by piecing together thousands of small ceramic tiles to make intricate mosaics. With the time-saving materials and methods available today, even much less talented—and patient—workers can produce an imaginative and attractive tile floor.

Since most small tiles come on sheets backed with plastic or mesh, the most practical patterns, such as accent stripes, are big and bold. But you can also cut sheets of tiles between rows to obtain narrow strips of tile for more intricate designs.

Even the larger quarry tiles and pavers can be used in varying colors and patterns to provide bold and unusual designs. Some striking examples are pictured on pages 22–27.

To create a pattern in a tile floor, sketch in the design on graph paper. As you establish working lines, transfer the design you've drawn to the floor. Complete instructions for working with ceramic tile begin on page 65.

Bringing concrete indoors

Though valued as a basic building material, concrete isn't often used for its inherent beauty. Yet special treatments can give concrete character and color. Used indoors, exposed concrete aggregate creates interesting patterns, is durable and waterproof, and provides a nonslip surface. See page 31 for an example.

Industrial options in carpeting

Floor coverings commonly used in offices, factories, and public buildings are selected because of their durability and relatively easy maintenance. For exactly the same reasons, many of these products are now finding their way into residential use in family rooms, bedrooms, kitchens, bathrooms, and hallways (see page 32 for an example). The products range from industrial or commercial-grade carpeting to studded or ribbed synthetic rubber flooring.

...Continued from page 6

natural colors. They're as rugged as quarry tile.

Patio tiles. Rough and irregular in appearance, patio tiles are thicker than quarry and paver tiles, but come in the same earth tones—tans, reds, and browns. Since patio tiles absorb water easily, they must be protected with a sealer. Tiles can shatter if frozen, so don't use them outside in cold climates.

Glazed tiles. Though usually glossy, glazed tiles can also have matte or textured surfaces which reduce the risk of slipping. They come in a range of colors, sizes, and patterns, from plain to fancy. Included in this group are the tiny mosaic tiles purchased already mounted in groups on plastic or mesh backing.

Masonry

Natural stone, such as slate, flagstone, marble, granite, and limestone, has been used as flooring for centuries. Today, its use is even more practical, thanks to the development of sealers and finishes.

Stone can be used in its natural shape or cut into uniform pieces. Manmade masonry products, specifically brick, are becoming increasingly popular for interior flooring. Because of its heat-retaining property, brick is suitable for use in passive solar heating designs. Even concrete, another manmade masonry product, can be used for flooring.

In addition to being easy to maintain, masonry flooring is good-looking, waterproof, and virtually indestructible.

The cost of most masonry flooring is high, mainly because of the expense of handling and shipping. Moreover, the weight of the material is such that a very strong, well-supported subfloor is required.

If you're considering a masonry floor, particularly one of natural stone, check your local building code and a building materials supplier; because of its weight, masonry isn't usually shipped great distances and tends to be distributed on a local basis.

Stone. You'll need a thick mortar base to lay a flooring of rough-hewn stone. But you can use thin-set mortar for stones that can be cut into pieces of uniform thicknesses, such as slate and marble.

Slate varies in color from dark blue to gray and green. If it's not available locally, slate can be a very expensive flooring, but it's hard-wearing, easy to maintain, and attractive. Like other masonry surfaces, it is fairly noisy and cold.

Marble, traditional and formal, comes in uniformly thin tiles that are relatively easy to install. Though durable and practically maintenance-free, marble is expensive, cold underfoot, and slippery.

Manmade products. Brick and concrete, once considered only outdoor materials, can be just as practical and effective indoors.

Bricks are made in full thicknesses and in "splits" (half as thick). Concrete slabs can take the weight of full-thickness bricks, but wooden subfloors require the lighter weight splits.

Concrete is, of course, used for slabs and is a perfect subfloor for almost any type of flooring. But concrete can also serve as a permanent floor, particularly if it's colored or painted. Exposed aggregate relieves the monotony of plain concrete and adds texture and traction.

Carpeting

Like resilient flooring, carpeting is available in an array of colors, styles, and materials, with prices that vary widely.

Comfort and appearance are two characteristics that have long made carpeting commonplace flooring for living rooms and bedrooms. But now even bathrooms and kitchens benefit from the development of hard-wearing carpeting materials.

Carpeting that has no backing is called "conventional carpeting" and requires a pad underneath. Carpeting with a bonded rubber packing is called "cushion-backed" and needs no separate pad.

You can install both types of carpeting directly over a well-prepared subfloor or over any type of old flooring that's clean, smooth, and free from moisture.

Carpeting options. Though natural fibers are still popular, synthetics dominate the market because of their lower prices and easier maintenance. Carpeting is available in wool, acrylic, nylon, polypropylene, and polyester, as well as combinations of these materials.

Half a dozen basic manufacturing processes are used to produce carpeting. Among the resulting styles are plush, shag, level loop, and sculptured pile.

Plush carpeting has a smooth, even surface that gives the carpet a soft look. After vacuuming, it will have a shadowed appearance. Because the pile is closely woven, the material doesn't crush easily.

Shag carpeting has a more casual look; it comes in solid colors, tweeds, or multicolored, and in fibers of different lengths. Not as dense as plush, shag carpeting is less expensive, but not as long-wearing.

Level loop carpeting is made of tightly constructed loops that make it easy to maintain and very durable.

Sculptured pile carpeting uses different lengths of pile to achieve varied surface levels. Rich surface texture and design characterize this type of carpeting.

WOOD FLOORS

WARMTH AND ELEGANCE FOR ANY ROOM

Simply Classic
Golden oak-plank flooring sets off this finely crafted home. Random length tongue-and-groove boards, blind-nailed with a nailing machine (see page 40), assure tight fit and hidden fastening. Staircase combines oak treads and handrail with cedar balusters and post. Architect: Ben Tarcher.

Patterns in Oak
Bold herringbone pattern began with oak-strip flooring cut to a constant length. Top-nailing made installation easy and provided control as work proceeded. Dark stain emphasizes the wood's grain, enriching the overall pattern. Design: Audrey M. Borland.

OPTIONS IN OAK

Appearances Can Be Deceiving
At first glance, this appears to be a random-plank floor. Actually it's made of more economical strip flooring. Plank effect was accomplished by the use of a router guided by a long straightedge. Routed grooves follow edges of oak strips at intervals of two to four strips. Architect: Peter C. Rodi/Designbank.

Planks in the Bath
Random-plank oak gives a warm feeling to this bathroom. V-grooved flooring features decorative plugs, simulating pegged installation. Polyurethane finish keeps away moisture. Design: Phillip Emminger.

Softly Shining
Translucent stain, lightly tinted, gives this floor subdued, elegant good looks. Stain minimizes grain, yet allows it to show through. Floor is blind-nailed, tongue-and-groove oak. Architect: Bert Tarayao.

On the Bias
Laid diagonally, grooved oak-plank flooring takes on additional interest. A single wide board, equal to the wall's thickness, marks the junction of two diagonals, helping to define the rooms. Dark stain accents the wood's natural grain. Design: Audrey M. Borland.

PARQUET POTPOURRI

Going with the Flow
Standard finger-pattern oak tiles lend continuity in this small split-level house. Painted risers effectively set off the steps without breaking up the flow of spaces. Architect: Ted T. Tanaka.

Elegance in the Kitchen
Oak tiles in classic finger basketweave pattern serve as a handsome foil for this kitchen's straight-lined European cabinetry. Easy installation makes this a practical choice for amateurs. Polyurethane protects oak's natural golden color. Design: Joan Simon.

Weaver's Craft
Square-end dividers are the secret of this vivid, yet simple, basketweave pattern. Large oak squares keep the pattern open, resulting in a true interwoven look. Design: George L. Stephenson.

Patterns in Gold
Oak parquet tiles in the traditional Mount Vernon pattern yield complex richness in this breakfast room floor. Each tile is composed of five strips contained in a mitered frame of the same strips. Put together, the tiles give the effect of squares surrounded by pointed pickets. Architect: Peter W. Behn.

QUARTET OF FLOORING WOODS

High School Revisited
Rescued from an old gymnasium, maple flooring now graces a country bathroom. Wood was remilled, then installed, sanded, and sealed with polyurethane. Architect: Ben Tarcher.

Two for the Price of One
Hemlock, washed with a light, translucent stain, creates a delightfully casual floor. This one carries a bonus: it's a ceiling, too. The 2 by 6 tongue-and-groove boards have been installed upside down, their V-grooved top sides forming a ceiling for the room below, their flat bottom sides flooring the master suite shown here. Architect: J. Alexander Riley.

Pine Light
Clear finish over floor's knotty pine boards helps keep the mood light and airy in this home where floors and walls are a near match. Blind-nailing yields an uninterrupted surface. Architect: Obie Bowman.

Pine Dark
Dark stain and satin finish give this pine floor a soft umber glow. Knots, as well as simple top-nailed installation, contribute to the room's country informality. Design: Curt Williams.

RESILIENT FLOORS

EASY-CARE ECONOMY

For a Carefree Kitchen
It's hard to beat sheet vinyl for easy care and minimum fuss. Restrained marblelike pattern of this no-wax kitchen floor blends easily with wallpaper and tile accents. Straightforward installation proceeds quickly. Design: HDL, Walnut Creek.

Octagons & Squares
Tile-look-alike floor features soft color, rich pattern, easy no-wax maintenance —all of which make it a far cry from yesterday's linoleum.

Team of Two
Bathroom flooring combination takes advantage of each material's strengths: sheet vinyl for easy care and comfort underfoot, ceramic tile for moisture protection in the plant bay. Tilelike vinyl pattern harmonizes with the real thing.

ARTFUL DESIGNS

On the Rocks
Water-washed stones inspired this vinyl's pattern. Its muted, natural tones blend with almost any decor, including this bright, modern interior.

On the Boards
Sheet vinyl gives a remarkable performance in the role of oak flooring, while contributing its own star qualities —fast installation, good looks, easy maintenance—in this remodeled kitchen. Architect: Pamela Seifert.

The Last Word
Custom-made white squares and midnight blue pickets combine in a piecework floor to dramatize the elegance of resilient vinyl. Each piece is set off by a deeply beveled edge. The cost: not low; the results: outstanding. Design: Audrey M. Borland.

PIECE BY PIECE

Factory Direct
Popular high-tech look draws inspiration from industrial design. Among the many products available for residential use, this black rubber floor tile is a standout; it's easy to install and tremendously rugged. Courtesy of Plus Kitchens.

Basic Brick
These vinyl tiles are ideal for do-it-yourselfers. Adhesive is already applied—you just peel away paper backing and position tiles on the prepared floor.

Tromp l'oeil
Strikingly realistic, "plank" flooring is actually resilient synthetic installed piece by piece.

CERAMIC TILE

DURABLE BEAUTY MORE THAN SKIN DEEP

Mediterranean Magic
Beautiful, practical, glazed quarry tile dresses up family room. Pattern is just one of an ever-changing variety on the market. Rugged glazed surface resists wear, makes setting tile easier since grout can't stain it during installation. Design: Color Tile.

Natural Look of Terra Cotta
Unglazed quarry tile on this floor gives a clean, crisp look to a room. At home both in the house and outdoors, it's best coated with clear masonry sealer for easy maintenance indoors. Though widely available in a variety of colors, shapes, and sizes, red-clay tiles are most typical. Architect: Peter W. Behn.

Kidproof
Equally resistant to the patter of little feet and the clomp of clodhoppers, glazed mosaic tile makes a practical—and beautiful—flooring for high-traffic areas. Design: Color Tile.

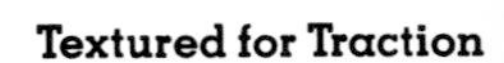

Textured for Traction
Though it's a preferred choice for high-traffic areas, glazed tile is often too slippery for safe use on stairs. These pavers overcome that problem by adding surface texture, achieving real beauty at the same time. Design: Franciscan Tile.

PATTERN VARIETY SIMPLY ACHIEVED

Borderline
A simple, yet distinctive, way to add emphasis with tile is shown in this bathroom floor detail. A line of small tiles forms a border, with bright corner touches. White tiles were cut to make room for color accent. Architects: Abrams, Millikan & Kent of Berkeley.

Square Root
Elegant geometry is created by this juxtaposition of two tile patterns. White tiles, divided into fourths by crisp blue lines, mate perfectly with the staggered array of blue tiles (divided into sixteenths by crisp white lines) marching resolutely to the tub. Architect: William B. Remick.

Circle Game
Plain and decorated tiles combine in circular patterns. A tile-cutting tool made the triangular and tapered pieces at centers of circles and in spaces between.
Design: George L. Stephenson.

Modular Meandering
Glazed modular tiles lend themselves to the attractively random pattern of this kitchen-sunroom floor. Each tile is a simple multiple or division of every other.
Architects: Jacobson/Silverstein/ Winslow.

MEXICAN PAVERS: THEME AND VARIATIONS

De la Casa al Patio
A true indoor-outdoor flooring, low-fired Mexican pavers are at home inside or on the patio. Pointed pickets add pattern variety. Indoors, pavers are coated with masonry sealer; outdoors, they're left bare. Architect: Edward Giddings.

Overseas Tribute
European tiles echo Mexican pavers shown above right, yet are different. Their perfect, matte-glazed surfaces are the result of sophisticated industrial manufacture; the Mexican originals are products of a handcraft that has changed little over centuries. Though the patterns are alike, the effects differ.
Design: MLA/Architects.

Serendipity
When solar greenhouse was added, Mexican pavers were a logical flooring choice. As a greenhouse floor, they take lots of abuse; as a passive-solar mass, they're ideal for soaking up sun by day and releasing it to the house by night.
Architect: Obie Bowman.

All through the House
Hexagonal Mexican pavers set up a pleasing overall pattern in this entry and living room. Square pavers simplify construction of steps. Architect: Peter C. Rodi/Designbank.

MASONRY FLOORS

STONE . . . FOR TIMELESS ELEGANCE

Sleek Slate
Black as midnight, cut slate makes a grand and gleaming entrance. Trimmed slabs are laid in mortar over wooden subflooring designed to carry the extra weight. Design: Judith Sutter.

Rugged Fieldstone
Roughly trimmed stones, laid in mortar over a concrete slab, lend solidity to this entry hall. Their installation required care and patience—both amply rewarded in the result. Design: MLA/Architects.

Sophisticated Marble
Marble makes a regal paving. In this sunroom, square slabs are laid in a formal grid that's appropriate to the material and easy to execute. Varying the grain direction of the stones adds subtle counterpoint. Design: Norman Roth.

OUTDOOR PAVING COMES INSIDE

Outside In
Brick patio surface becomes the floor of an added-on sunroom. Enclosing a section of patio like this makes both spaces work harder; the house gains a new room, the patio gains extra shelter. Brick floor is a practical choice for both. Landscape architect: Woody Dike.

Patio Recycled
Fresh appearance of these bricks belies their age—they spent more than 30 years outdoors before coming inside. Salvaged from walks and patios torn up during construction of a new house, they now make a practical, informal floor that's part of the home's solar design. Architect: Rob Wellington Quigley.

New Wrinkle
This imaginative floor is made up of tinted stonelike concrete pavers cast in molds lined with plastic sheeting. Inverted and mortared into place, the pavers show the imprint of wrinkled plastic, yet are smooth and finished-looking. Design: Joan Simon.

Concrete Comes Home
Seeded-aggregate concrete, coated with masonry sealer, makes a strikingly beautiful, economical, low-maintenance floor. It can be poured as part of a foundation slab, or on top of one. It's as practical as plain concrete, and more attractive. Architect: Richard Pennington.

CARPETING

CONVENIENCE BY THE YARD

Corporate Contribution
Tightly woven commercial carpeting, formerly reserved for use in the workplace, is finding increasing home usage. Its easy maintenance and extreme durability make it ideal for a teenager's bedroom. Architect: William B. Remick.

Perennial Favorite
Cut-pile wall-to-wall carpeting makes a good background for almost any decor or architectural style. Recent developments in simplifying installation make it possible to do it yourself.
Design: Barbara Elliott.

Breaking New Ground
Short plush carpeting, incorporating an impermeable layer to resist stains and make cleaning easier, is now available for bathrooms and kitchens. Muted plaid of this carpet makes it right at home with natural wood cabinets.
Design: Centex Homes.

SOMETHING EXTRA

Custom Contours
Bold, neat border was custom fitted to the room, effectively dressing up plain carpet and tying together key colors. It's simpler than it looks: many large carpet outlets can make up borders like this one, using standard seaming techniques to combine varying colors of the same carpeting. Design: Rex Carpets.

Tribal Origins
Nubby, loop-pile carpets like this one are loosely termed "Berber" carpeting. Most are available only in wool, but synthetics, with considerably lower prices, are increasingly easy to find. This room takes its cue from the carpet's natural colors; earthy Berber designs are equally good with bright-hued color schemes. Design: Steven Chase.

Surface Variations
Modern counterpart of "sculptured" rugs, this carpet is enriched by a simple, incised geometric design. Design: Edward Field.

SPECIAL EFFECTS

PRACTICAL COMBINATIONS

Island of Slate
One neat route to an attractive wood-stove hearth is a slate panel that sits directly on subflooring. The slate was selected for grain and color pattern and for an exact match to the thickness of the surrounding oak. Design: Russ Williams.

Directing Traffic
In this living room, a mix of materials defines the function of different areas—oak parquet tile for traffic, and a dropped-in sweep of carpet for the seating area. Curving border of oak-strip flooring marks the junction. Design: William Young.

Natural Combination
Redwood headers divide this exposed-aggregate floor into sections for visual interest and building ease. The floor was poured early in house construction, allowing the redwood to collect a weathered patina as workers trooped over it daily in rainy weather. Later, it was simple to scrub the floor and apply a sealer. Architect: J. Alexander Riley.

Unusual Alliance
Sections of particle board divided by redwood strips set with terra cotta tiles make up this attractive, economical floor. Polyurethane keeps it shining. Architects: Jacobson/Silverstein/ Winslow.

...SPECIAL EFFECTS

VISUAL DELIGHTS

Picture Perfect
Look closely: it's not a rug. In colonial days, those who couldn't afford rugs simply painted them on the floor. Today, this craft is being revived, making a new virtue out of an old necessity. Design: Liddy Schmidt, Linda Kilgore.

Gleaming Wrap-up
Torn brown wrapping paper covers the concrete slab floor of this artist's studio. High-gloss polyurethane provides protection and makes for easy care. The resulting combination provides all the utility of concrete, but with striking visual uniqueness. Architect: Ron Yeo.

Colorful Overhaul
This old floor underwent a facelift. After sanding, a green glaze was applied with a cloth, and a painted white border was added. Stencils followed, in white and green. Finally, came several polyurethane coats to protect the careful work. Design: Liddy Schmidt, Linda Kilgore.

Sea Change
Suggestive of caulked boat decking, this elegant floor resulted when its owner stripped a dark stain and substituted a bleached finish. "Caulking" is actually acrylic strips set between groups of oak planks. Design: MLA/Architects.

CAUTION

Tools of the Trade
Just strike the plunger and the nailer does the work, cinching tongue-and-groove boards together and driving a nail through the tongue. The result: a perfect, squeak-free floor.

INSTALLING FLOORING MATERIALS

Step-by-step instructions on what to do and how to go about it

- wood—still a popular choice
- resilient—easy to work with
- ceramic—planning is crucial
- masonry—bulky but worth the effort
- carpeting—for classic comfort

No matter what type of floor you want for a new home or a remodeling project, you can be sure the manufacturers of popular flooring materials are doing everything possible to simplify the installation of their products. Your building materials supplier or flooring contractor should also be able to help you complete a flooring project successfully, supplying you with installation instructions and suggestions as to the proper tools and materials you'll need to do a good job.

Innovations in design and manufacture have made many flooring materials easier to install and less taxing on the homeowner's skills. For example, ceramic tile, once thought to require the skills of a professional, can be installed today by anyone with patience and average building talent. Intricate mosaic patterns come in sheets with a backing of plastic, cotton mesh, or paper that makes the tiles easy to handle. Tile suppliers will help you choose the proper adhesives, grouts, and tools.

Hardwood strip and plank flooring and wood block flooring can be purchased with durable factory-applied finishes. Wood tile is available with adhesive backing.

New types of carpeting materials and simpler methods of installation make laying some carpeting a relatively easy job for today's homeowner. Cushion-backed carpeting can be installed without stretching; seams in synthetic conventional carpeting can be ironed together with hot-melt seaming tape.

Vinyl tiles and other resilient tile flooring also come with adhesive backing or can be set in easy-to-apply adhesives.

In this section you'll be shown how to install all the basic traditional flooring materials. You'll discover that many of the more popular materials are among the easiest to work with.

Is it a job for you?

Though many kinds of flooring are now designed to make installation easy for the do-it-yourselfer, there are still, of course, projects best left to professionals. If you're thinking of putting in your own floor, read through the installation instructions that follow before making a final choice of materials. Make sure you're looking at flooring that is not only suitable for the room where you plan to put it, but also installable by someone with your experience and skills.

Don't overlook the fact that preparing a suitable base for new flooring can be more complicated and time-consuming than laying the final floor. Remember that the type of materials you choose will affect the amount of preparation necessary; for example, ceramic tile, because it's inflexible, requires a particularly stable subfloor. Finally, keep in mind that not all types of flooring are suitable for all kinds of conditions; wood, for example, which is affected by moisture, should be considered for rooms below grade only if precautions are taken to control moisture.

After weighing the advantages and disadvantages of the many types of available flooring, you'll probably find that you can limit your selection to the most practical materials and still achieve the appearance and quality you want.

Is the structure in good shape?

Before you undertake the installation of any new flooring—preferably before you even order the materials—take time to check the structure of the floor to make sure it's in good condition. The illustration at right shows how a floor and its supporting structure are put together in a typical frame home. For a guide to making a thorough inspection, see "Locating the cause," pages 87–92. This is a very important step, one that could save you time and money in the future.

WOOD—STILL A POPULAR CHOICE

A traditional flooring material for American homes, wood was widely used in bygone days, because it was abundant and inexpensive. Though good wood flooring is still readily available, it's no longer low in cost.

Yet wood has remained popular because of its warm, natural look and its resiliency and long life—advantages that can offset its high original cost. A good wood floor will last the lifetime of most homes, can be refinished several times, and will actually improve with age.

Wood flooring may be bought with a factory-applied finish, or unfinished for sanding and finishing in place.

Red and white oaks are the most common species for flooring, but other hardwoods are available, including teak, beech, birch, hard maple, pecan and more exotic species. Some softwoods, such as fir and pine, are also used for flooring.

THREE BASIC TYPES

Wood flooring is milled in many different shapes and can be laid in an endless variety of patterns. But there are three basic types of wood floors: strip, plank, and block. The first two actually contribute to the structural strength of a house; most wood block flooring, on the other hand, is considered a floor covering only. Wood block flooring may be solid or laminated of several layers. Only laminated flooring should be installed on below-grade concrete subfloors, and then only if precautions are taken to keep the concrete dry (see page 47).

Strip and plank flooring are rated according to quality. Color, grain, and such imperfections as knots are assessed to determine the grade. The best grade for both is "Clear," followed by "Select," "No. 1 Common," and "No. 2 Common." Other hardwoods have similar grading systems, but with different designations; check with your dealer.

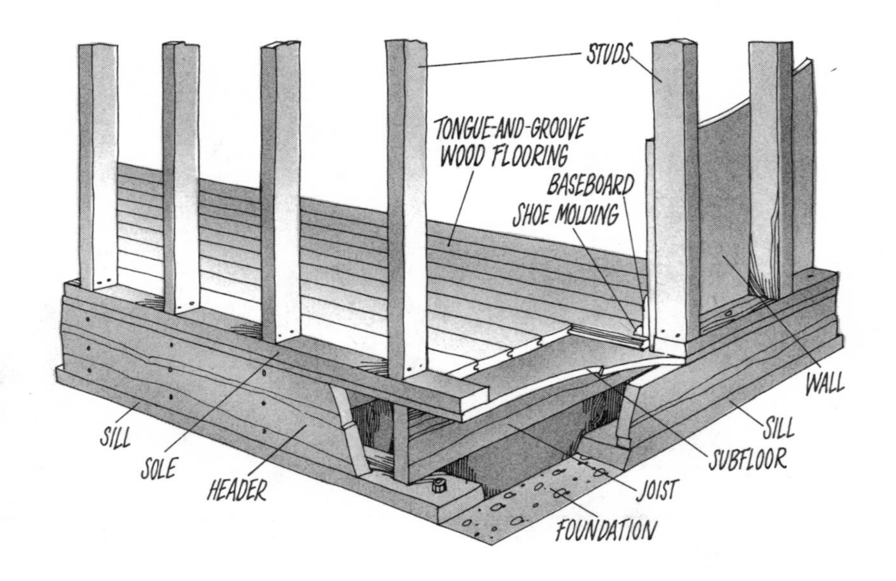

Strip flooring

This is the basic hardwood flooring, made up of narrow boards with tongue-and-groove edges and ends laid in random lengths.

The most commonly used strip flooring for finishing in place is ¾ or 25/32 inch thick, with a face width of 2¼ inches, though you can buy widths that vary from 1½ to 3¼ inches. This flooring is suitable for most residential use. Thinner strips (5/16, ⅜, and ½ inch) are available with either tongue-and-groove or square edges. Thicker boards are available, but are usually installed only in commercial buildings. Tongue-and-groove boards are usually end-matched, with a tongue at one end and a groove at the other end of each piece.

Strip flooring comes in random lengths, with individual boards usually 2 to 8 feet long; they are grooved on the back to give the floor resiliency and to make it easier to lay boards over minor subfloor irregularities.

Plank flooring

A holdover from colonial days, plank flooring comes in random widths as well as random lengths. Most plank flooring sold today differs little from standard strip wood flooring. The individual boards are milled with tongue-and-groove edges and ends, or square edges, and can be installed in the same way as strip flooring. Or plank flooring may be laid in adhesive, just as wood blocks are (see page 55).

The major difference between conventional strip flooring and plank flooring is that planks are produced in random widths (usually about 3½, 5½, and 7½ inches) as well as random lengths. Plank flooring may be ¾ or ⅜ inch thick and either solid wood or laminated—the latter better for below-grade installations.

Generally, if the planks are no more than 4 inches wide, they can be installed exactly like strip flooring. But if planks exceed 4 inches in width, they should be installed over a plywood subfloor; if you're working over a base of screeds, they should be covered with a plywood subfloor before the flooring is attached. Instructions for both kinds of subfloor preparation appear on pages 47 and 48.

Plank flooring is available prefinished, either with real plugs or with simulated ones.

Block flooring

The terms parquet, wood mosaic, and wood tile are used almost interchangeably to describe wood flooring laid in blocks or squares. All are types of wood block flooring. The "blocks" may be solid wood pieces, laminated sheets, or squares assembled from smaller wood pieces (generally called parquet).

Wood block flooring has become increasingly popular because of the wide selection of styles and the ease of installation. Improved, easy-to-install materials now make it possible for a homeowner with average building skills to create attractive, quality wood block floors.

Block flooring is most commonly manufactured in 6 to 12-inch squares, but it's also available in rectangles or in panels up to 39 inches square and usually 5/16 to 13/16 inches thick, held together with mesh backing.

Blocks are often called wood tile because they are usually laid in mastic, much as resilient tile is installed. Some thick blocks have tongue-and-groove edges and may either be laid in mastic or nailed through the tongues; the latter is a job for professionals. Parquet or mosaic blocks are usually made up of many smaller pieces of wood bonded to a single piece of wood backing or held together with cotton or plastic mesh.

Laminated tiles—some as thin as ⅛ inch, but most of them 5/16 or ⅜ inch—are produced with a surface of hardwood veneer. They may be backed with mesh, and they're laid in adhesive. Such wood tiles are also available with an adhesive backing that makes it possible to set them directly on a subfloor. The manufacturer will provide specific instructions on how to prepare a subfloor for such flooring. Laminated blocks are the best choice for below-grade installations.

Solid block squares are made of short lengths of wood held together with splines of metal, wood, or plastic. This type of block flooring can be purchased in thicknesses from 5/16 inch to ¾ inch and more.

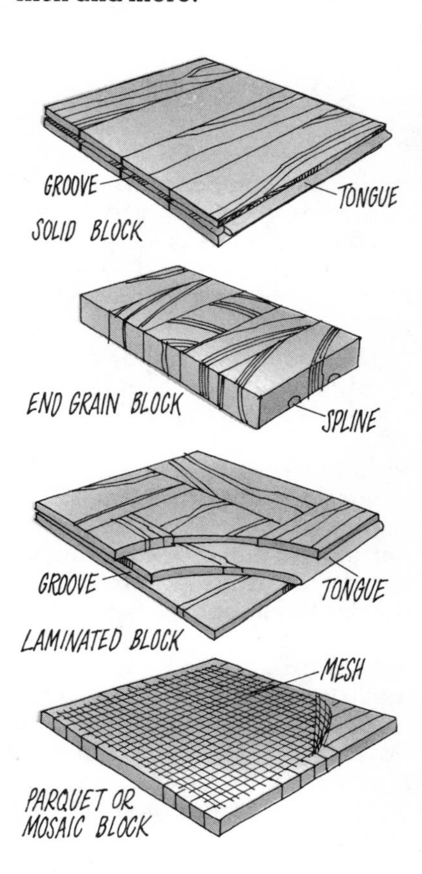

Another type—end grain block flooring—is exceptionally durable and has long been a popular choice for use in commercial buildings. But its appearance and long life make it practical for use in homes as well. This type of block flooring ranges from 1 to 2½ inches thick and can be bought in squares or in rectangles as long as 18 inches.

Blocks may be made with square or tongue-and-groove edges.

Before selecting a specific type of wood block flooring for your home, take time to investigate your options. Block flooring comes in virtually every kind of wood in a seemingly endless number of patterns. You can buy it with a factory-applied finish, or sand and finish it after installation. Examples of wood block flooring can be seen in the color section beginning on page 12.

IS WOOD THE RIGHT CHOICE FOR YOU?

In the following pages, you'll find information on installing the three basic types of wood flooring—strip, plank, and wood block—none of which is overly difficult to work with.

But before you settle on wood flooring, consider the following two notes of caution.

Preparing a proper base for any of the three kinds of wood flooring can be more demanding than putting in the new flooring itself. For details, see "Preparing the Subfloor," pages 46–50.

Moisture, the enemy of wood, can make some rooms unfit for wood flooring. In particular, wood floors are seldom installed in rooms below grade or in areas subject to dampness, unless laminated flooring is used and careful steps are taken to keep moisture from penetrating the slab and reaching the wood. It's important to check an on-grade concrete slab carefully for moisture before installing wood flooring over it (a simple moisture test is described on page 46), and you must be sure it will stay dry throughout the years—otherwise you'll have to seal off the moisture from the wood (see page 47). Similarly, the space below a standard floor supported by joists and beams should be properly ventilated and protected from moisture if wood flooring is to be laid over it.

ORDERING AND STORING YOUR WOOD FLOORING

Any flooring materials supplier will be able to tell you the exact quantity and cost of the flooring you'll need if you provide the exact measurements of the area, or preferably, a scale drawing.

Order enough flooring to allow for waste and to leave yourself some extra pieces to put away for future repairs.

Wood flooring needs time to adjust to conditions in the room where it's to be installed, so it's important to plan well ahead. When you order, discuss the date of delivery with your dealer and make sure you'll be ready to receive the materials and store them properly.

If the manufacturer and supplier have done their jobs well, your flooring will be delivered properly dried. If it's allowed to absorb moisture at the building site, it will shrink and show cracks after it has been installed in a warm, dry room. So it should never be delivered in rain or snow unless well protected, and it should not be stored outside or in a cold or damp building.

If you're dealing with a new house or room addition, have all doors and windows in place and the structure closed in before the flooring is delivered; wet plaster or masonry, which releases moisture into the air, should be allowed to dry thoroughly.

New, unwrapped strip or plank flooring should be delivered at least 4 days before installation and stored, if possible, in the room in which it will be put down—or in any case in a warm, dry environment. The temperature and humidity should be close to levels that will be normal for the flooring's new home. Untie the bundles and stack the individual boards loosely (see drawing) so air can circulate through the stacks.

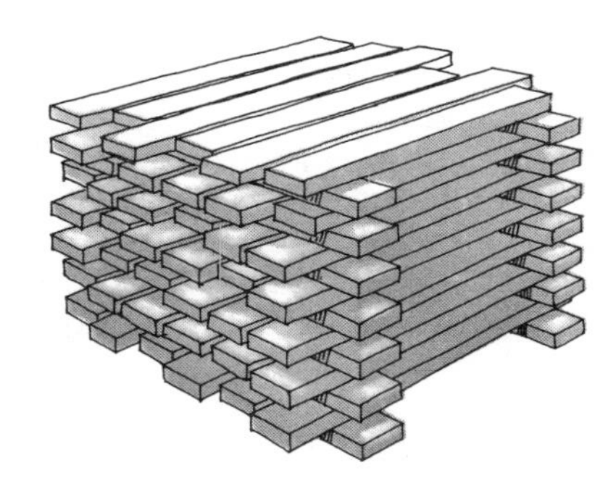

Wrapped flooring will have been packaged and sealed under ideal conditions and should be ready to install when it's unwrapped.

Wood block flooring should be treated in the same basic way as new strip or plank flooring when delivered. Plan on unpacking it and storing it where it's to be installed, at least 72 hours before beginning installation, but not until the humidity and temperature are approximately the same as they'll be when the room is occupied.

Some block flooring, particularly prefinished flooring, is packed in sealed cartons. Check the manufacturer's instructions before unpacking it.

When you unpack new wood block flooring, stack it loosely so that air can circulate around the individual pieces.

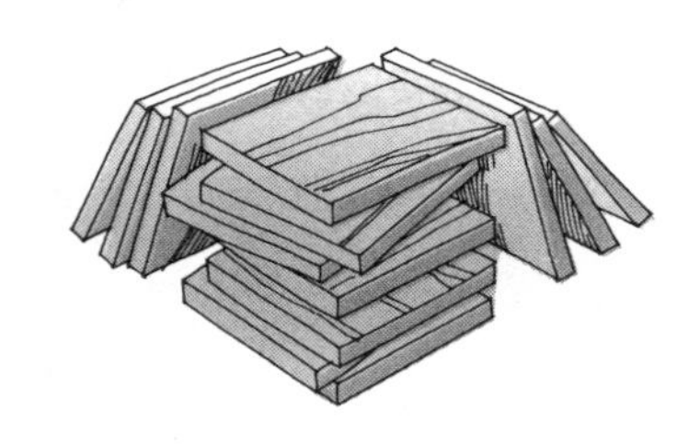

TOOLS AND SUPPLIES

The tools and supplies you'll need will vary with the type of wood flooring you plan to install. The following lists cover what's needed for the actual installation of each type. You may need additional tools and materials to prepare a suitable subfloor, so before heading for the hardware store or lumberyard, read "Preparing the Subfloor," pages 46–50.

With strip flooring (or unpegged tongue-and-groove planks), you can do a perfectly acceptable installation using basic hand tools. But the key to speeding up the task is a nailer, available from most tool rental companies (see illustration below). Special nails known as cleats feed automatically into the nailer and are driven at the correct angle down through each board to secure the floor. The nailer has a spring-operated mechanism that's triggered by striking with a rubber mallet. It is not difficult to operate and after a little practice on a few pieces of scrap flooring, you should be able to master it with little trouble. (Square-edged flooring is face nailed by hand.)

You'll also need a standard claw hammer, a nailset, a crowbar, a carpenter's square, wood chisel, a flexible measuring tape, a chalk line, a hand drill or portable electric drill with bits, and handsaws. A ripsaw or portable circular saw will be needed for cutting boards with the grain, and a back saw and miter box for cutting boards across the grain. If you have access to a table saw or radial-arm saw, you can cut hardwood flooring with considerably less effort.

If you're working over a concrete slab subfloor, you'll need a paintbrush to apply sealer and a notched trowel to spread mastic when preparing a wood nailing base for the new floor; follow the mastic manufacturer's recommendations as to the correct size trowel.

You'll need a supply of nails for blind nailing tongue-and-groove boards. Even if you're using a nailer (which uses special cleats or nails), you'll need a supply to nail those areas where the nailer cannot be used.

The nails are sized according to the thickness of the flooring—7 or 8-penny annular ring or cement-coated nails for ¾-inch-thick material, 5-penny annular ring or casing nails for ½-inch-thick material, and 4-penny casing nails for 5/16 or ⅜-inch-thick material. Your flooring supplier can help you select the right nails for the flooring you buy.

You'll have to face nail some of the boards and set the nails, so to fill these holes you'll need some wood putty that matches the color of the boards.

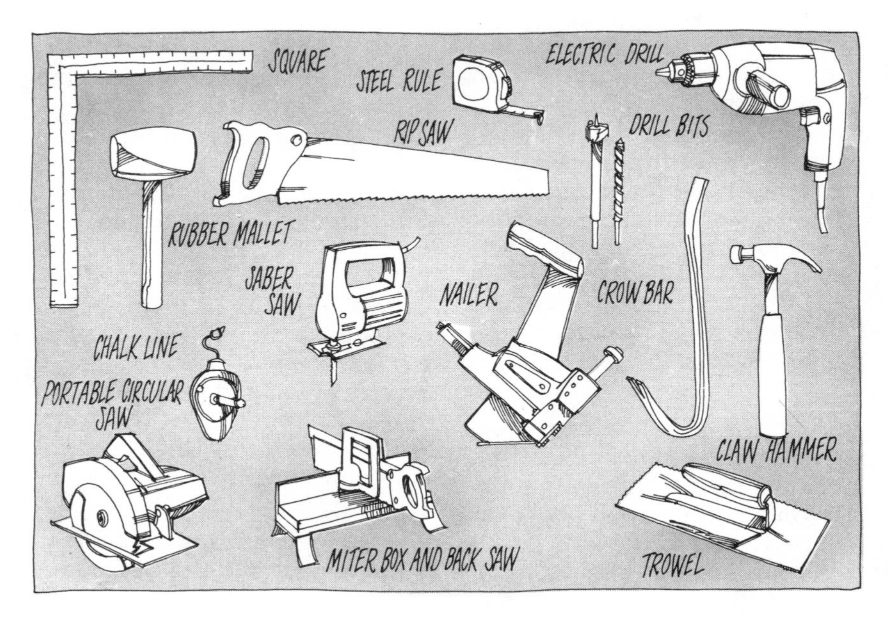

For pegged plank flooring, you'll need the basic equipment necessary for installing wood strip flooring (see previous section), and you may also need some additional tools if the planks you buy aren't predrilled for screws and plugs. If you do have to drill these holes, the preferred method is to counterbore the plug holes with a power or brad point bit, then drill the countersink and the clearance and pilot holes for the screws with a combination bit for the size screw you use.

If you are very careful to drill straight and hold the drill steady, you can drill the counterbores with a spade bit. You can also use four individual bits to make the counterbore, countersink, and clearance and pilot holes, but each hole will require four separate operations instead of only two.

The bit sizes depend on the screw sizes. If you can borrow or rent a second portable electric drill with a screwdriver attachment, you'll be able to save time by not having to change bits frequently. You won't need these additional tools if you plan to lay the flooring in adhesive or nail through the tongues.

Depending on the thickness of the planks you'll be installing, you'll need #6, #9, or #12 flathead woodscrews; the thickness of the planks and subfloor will determine the length of the screws. Your flooring supplier will be able to recommend the correct size screw for the planks you've purchased.

(Continued on next page)

...Continued from page 45

Finally, you'll need either precut wood pegs—often supplied with plank flooring—or hardwood dowels from which to cut your own pegs.

For block flooring, few specialty tools are required. You'll need a claw hammer, crosscut saw, putty knife, square, and steel tape. An electric saber saw is ideal for cutting tiles to fit around obstructions. A rubber mallet is handy for tapping block flooring into place without marring the wood surfaces.

To attach wood block flooring, you'll need an adhesive and possibly a primer for the surface to be covered. The flooring manufacturer will suggest adhesives that will work well with your type of flooring. The kind of adhesive may vary with the type of subfloor to be covered. The adhesive manufacturer, in turn, may recommend that you prime the surface before applying adhesive.

You'll probably have to buy a special notched trowel to apply adhesive; the flooring manufacturer will recommend the specific trowel for the adhesive recommended. And don't forget to buy the thinner or solvent recommended on the adhesive container to clean up spills and smudges.

Some manufacturers recommend that you use a heavy floor roller to seat flooring in adhesive. A roller (available from most tool rental shops) is particularly helpful if you have a large area to cover. To avoid damage to the floor's surface, be sure the roller is covered with a resilient material that can be kept clean.

If you install unfinished wood block flooring, you'll have to sand the floor with a floor sander and finish it with a protective sealer and finish. You can rent a floor sander from your flooring dealer; for tips on operating it, see page 99. Your dealer can help you select the right sealer and finish.

PREPARING THE SUBFLOOR

Whichever of the three types of wood flooring you plan to use, preparing a reliable base requires the same basic steps. Wood floors are typically laid over a concrete slab, over a wood subfloor supported by joists and beams, or in some cases over an existing floor—depending on the old floor's composition and condition, and on the kind of wood flooring you plan to install.

Block flooring, of course, requires a solid, continuous subfloor, whether of boards or plywood panels. Laminated wood tile can be laid directly in mastic on a *thoroughly* dry concrete slab that has a waterproofing membrane below or above to keep it dry. But blocks of solid wood pieces are best installed over a plywood base built over the slab.

This section includes the information you'll need to prepare a concrete slab, a wood subfloor, or an existing floor for new wood flooring.

Preparing a concrete slab

Before installing any kind of wood floor over a concrete slab, you must make sure the slab is *dry, level,* and *clean.* Even if a below-grade or on-grade slab appears dry, applying a vapor barrier under the floor will safeguard against future moisture problems. In addition, pay close attention to the recommendations that accompany the flooring you're installing. Each supplier will have specific guidelines for slab preparation and adhesives.

Checking for moisture. It's essential that the slab be free from moisture for an entire year before you install a wood floor over it. You'll need to observe it and check for moisture periodically, particularly during wet times of the year. Watch for moisture not only from below but also from other sources—seepage through a wall or condensation dripping from a pipe—that could ruin your new floor.

If you're thinking of installing a wood floor over a recently poured concrete floor, this year of observation should allow ample time for the concrete to cure. Just make sure the area is well ventilated, and turn the heat on during the cool season.

A simple test will tell you if a slab, new or old, is still releasing moisture. Cut a 16 to 20-inch square of clear plastic, tape the square to the concrete floor (sealing the edges with tape), and leave it in place for 2 or 3 days. Then lift the plastic and see if any moisture has condensed on the underside. If it has, give the floor more time to dry.

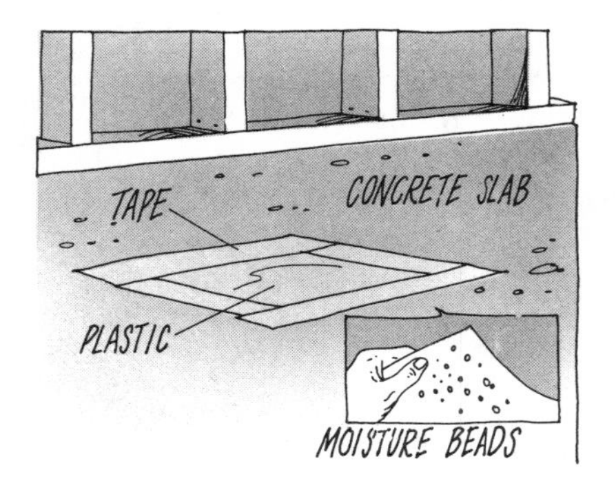

If a slab has had ample time to dry but is still damp, it's time to check the downspouts (consider extending them away from the house), assess the general drainage around the house, and look for leaky pipes. If improving drainage or fixing leaky plumbing doesn't correct a moisture problem and you cannot be sure that the floor can be kept dry, you should consider a type of flooring other than wood—one that will be less affected by moisture.

Cleaning a slab. Once you're satisfied the slab is dry, make sure it's level and clean. Small blobs of mortar or other spills of construction materials should be chipped away with a cold chisel. Fill minor dips or other irregularities with a patching compound.

Sweep the slab clean to remove as much dirt and dust as possible. *Do not clean the floor with water.* Most auto supply stores sell a chemical cleaner that will remove grease and oil.

Finally, to provide rudimentary protection from moisture, brush or trowel on a coat of asphalt primer directly over the slab, and allow it to dry thoroughly.

Provide a moisture barrier. Your next step will depend on the type of moisture barrier you intend to use. A single layer of polyethylene film laid just under new wood flooring is usually considered adequate moisture protection for a dry slab floor (see instructions in the next section, "How to prepare a base for wood flooring over a slab"). But if you suspect excess moisture might be a problem, or if the slab is on or below grade, you'll want extra moisture protection. A "two-membrane" vapor barrier laid directly over the primed slab can give you this protection.

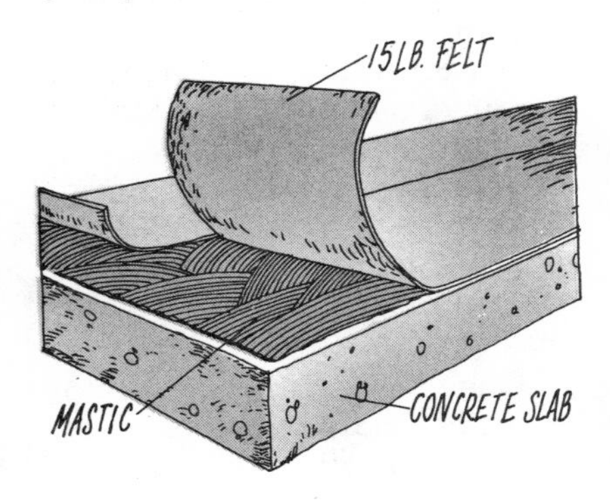

To install such a two-membrane barrier, use a notched trowel to spread a coat of asphalt mastic over the primer. Allow the mastic to dry. Then roll out a layer of 15-pound asphalt-saturated felt, butting the edges and ends of each course.

Trowel on a second coat of mastic, then roll out a second layer of asphalt paper so that the seams run parallel to, but fall between, the seams of the first layer.

How to prepare a base for wood flooring over a slab

Except for some kinds of laminated tile, wood flooring cannot be secured directly to a concrete slab. That means you'll have to lay either strips of wood (called screeds or sleepers) or a base of ¾-inch exterior plywood over the slab so the flooring can be attached to the wood. Screeds will work only for strip or plank floors; wood block flooring requires a solid, continuous base of plywood.

Using screeds. The best material to use for screeds is 2 by 4 lumber, pressure treated with chemicals for pest and moisture resistance. If you've installed a two-membrane moisture barrier, you can set screeds in mastic directly on the asphalt paper. If not, set them in a coat of hot asphalt mastic applied directly over the primer, or attach with lag bolts, depending on what type of screeds you choose.

Two ways to lay screeds. There are two common ways of laying screeds—in continuous strips or in staggered, short lengths (see illustration below). Continuous strips are the best choice in smaller rooms and over uneven concrete slabs. You can attach them with lag bolts secured to lead anchors set in the concrete slab, and adjust them with shims to provide a level base for your new flooring.

But staggered screeds are simpler to install, as the pieces—only 18 to 48 inches long—are easy to handle and provide air circulation under the floor. These shorter screeds are imbedded in mastic on 12 to 16-inch centers. They're called "staggered" because they're overlapped at least 4 inches where they meet.

Staggered or continuous, screeds should be installed at a right angle to the direction you plan to lay your strip or plank flooring. No screeds should be placed closer than half an inch to any wall, to allow for air circulation. You'll need to vent the space between the screeds to the room. You can do this with at least two openings cut through the floor close to the two walls at the ends of the screeds. Cover the openings with warm-air registers or ventilating louvers.

Check the level of screeds. When the screeds are in place, take a long straightedge (the uncut edge of a sheet of ¾-inch plywood works well) and check to see that the screeds are level. Plane down high points. If the screeds are attached with lag bolts, you can build up low areas with wood shims (wood shingles work well) placed under the screeds.

(Continued on next page)

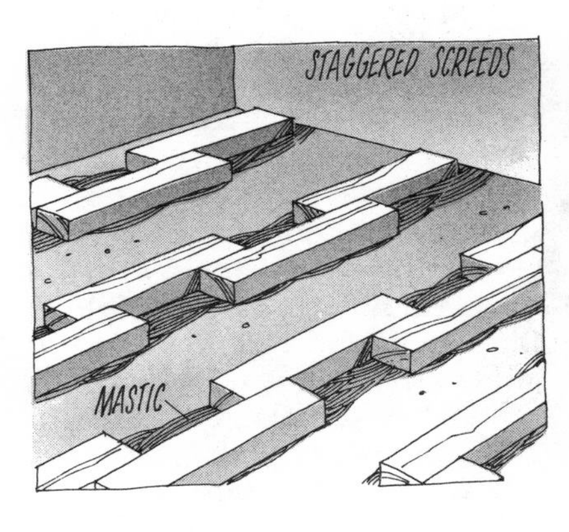

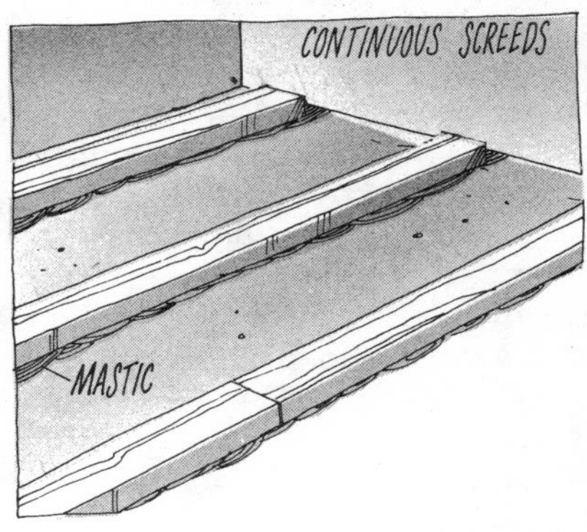

...Continued from page 47

Insulation is optional. If you want additional moisture protection, a quieter floor, and some extra insulation, now's the time to pour vermiculite or perlite insulation between the screeds and level it with the top of the screeds.

Cover screeds with film. After leveling the screeds, lay 4 or 6-mil polyethylene film over the top of the screeds for moisture protection. Overlap the edges of adjacent sheets of film at least 6 inches; it doesn't have to be attached. Take care not to puncture it.

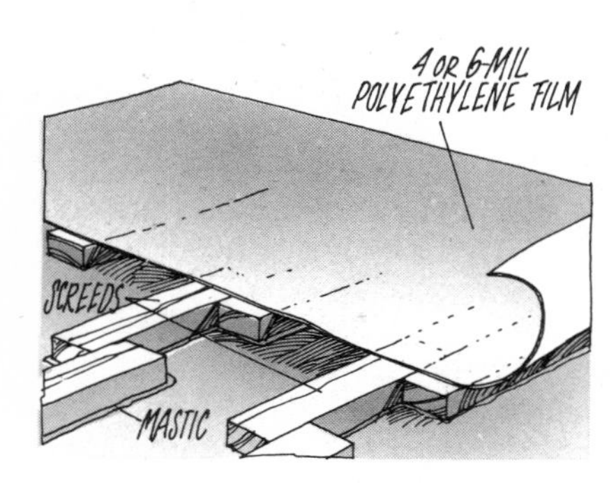

You now have a good base for your strip or plank flooring. If installing a wood block floor, nail ¾-inch exterior grade plywood to the screeds. Use 6-penny cement-coated or ring-shank nails and stagger the joints. Leave gaps between the plywood and the walls and doorways, as described below.

Using plywood. To prepare a solid plywood nailing base over a concrete slab, first make sure the slab is dry and clean, as outlined in the preceding discussion on testing for moisture and cleaning a concrete slab. If you haven't already done so, seal the dry slab with a coat of asphalt primer. Then cover it with a layer of 4 or 6-mil polyethylene film (no mastic is required); overlap adjacent sheets of film 4 to 6 inches and extend the film under the baseboards on all sides of the room (for tips on removing molding and baseboards, see the facing page).

Lay out panels of ¾-inch exterior grade plywood over the entire floor, cutting the first sheet of every other run so that end joints will be staggered (see illustration). Leave a space of ¼ to ½ inch between panels and the walls. Around doors and other obstructions, where gaps will not be hidden with molding, cut the plywood to fit, leaving a gap of only about ⅛ inch; the flooring will hide these spaces.

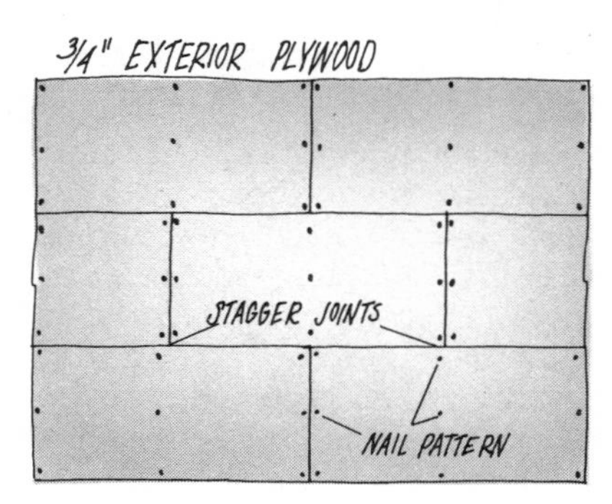

Starting at the center of each panel, use concrete nails to secure the plywood to the slab. Use at least nine nails per panel, in the pattern shown in the illustration above. You'll attach your new wood flooring directly to this plywood base.

If you're building a new subfloor

Either plywood or common pine or fir 1 by 4s or 1 by 6s provide a suitable base for wood flooring, and if you're putting wood flooring in a new home or addition—that is, if you can choose the sort of subfloor you'll be working with—those are your choices. Because it won't warp, ¾-inch plywood is generally considered the best subfloor material; it's easier to work with, too.

Lay plywood sheets with the grain running at a right angle to the joists, and nail them to the joists about every 6 inches. Butt panel ends together over joists—this should be no problem if the joists are on standard 16 or 24-inch centers—and stagger the joints in adjacent runs (see drawing).

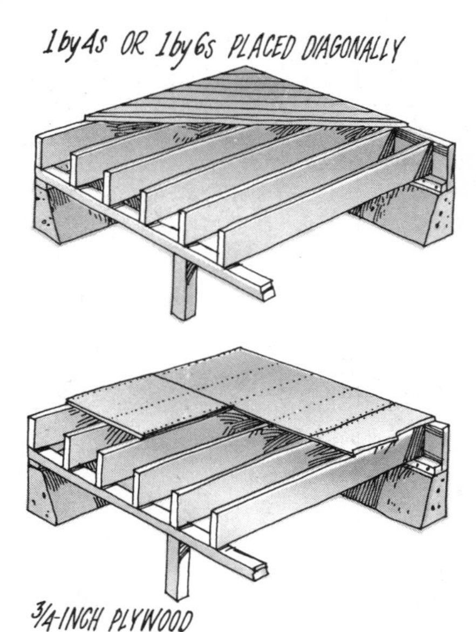

Lay boards diagonally across the joists and nail them in place (use two nails where each board crosses a joist), with 1⁄16-inch spacing between boards to allow for expansion.

Preparing an old floor for wood flooring

Though it's possible to lay wood flooring over an old floor that's in good condition, you may need to remove the old flooring to get down to the basic subfloor and make any necessary repairs or install underlayment. In the long run, this usually provides the most reliable base for your new floor.

Check the exposed subfloor for loose boards or loose plywood panels. If planks are too badly bowed and cannot be flattened by nailing, give the floor a rough sanding with a commercial floor sander (see "Tools and Supplies," page 97) or cover it with ⅜ or ½-inch plywood or particle board.

Should you have an irregular subfloor made of planks less

than 4 inches wide, and you intend to install wood block, cover the old planks with a ¼-inch layer of untempered hardboard, rough side up, or ¼-inch plywood. For planks from 4 to 6 inches wide, use ⅜-inch plywood; for planks over 6 inches wide, ⅝-inch plywood is best.

Fasten down ¼ or ⅜-inch material with 3-penny ring-shank or cement-coated nails; for ⅝-inch plywood, use 4-penny ring-shank or 5-penny cement-coated nails. Space the nails 6 inches apart across the surface of the panels.

However, an existing wood floor in good condition can, if you wish, serve as a base for new strip, plank, or block flooring; in fact, block flooring can even be installed over old resilient sheet or tile floor covering. The advantages of laying new flooring over old are that you bypass the messy job of removing the old flooring and you gain some soundproofing and insulation from the old floor.

One disadvantage in leaving old flooring in place is that you'll be unable to inspect the subfloor and correct irregularities. Another disadvantage is that there may not be enough space above appliances (a dishwasher, particularly) after the new floor is installed. And if you don't put new flooring under the appliances, you may even find it impossible to remove them in the future.

Whatever kind of old flooring you're faced with, begin your preparations by removing all doors, grates, and other obstructions in the room. Use a chisel to take up the shoe molding from around the baseboards, taking care not to split the wood (see drawing above). Remove the baseboards themselves, avoiding damage to the walls or door frames. As you take up molding or baseboards, number the pieces with chalk or a pencil so you can replace them in their

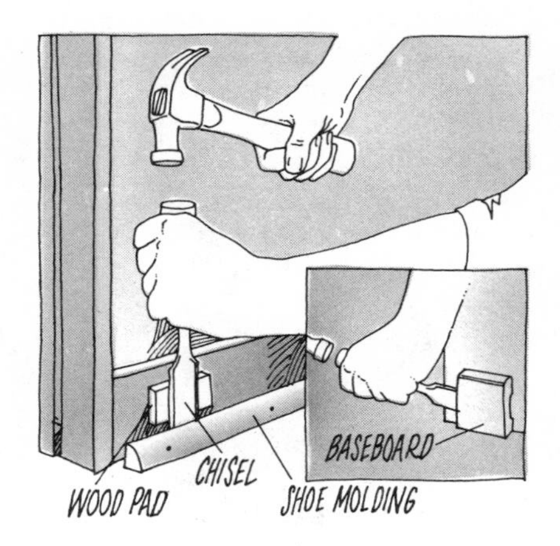

original positions once your new flooring has been installed.

An old wood floor must be structurally sound and perfectly level before you put new wood flooring over it. Separated, bowed, or buckling boards or any evidence of moisture damage in the original floor should serve as warnings of problems that you can't hide—for long—simply by covering them with new flooring. If you have any reason to suspect the old floor or its supporting structure, examine it very carefully before installing a new floor over it. (See "Locating the cause," page 87; beginning on page 88, you'll find suggested remedies for the most common structural problems.)

Once you're satisfied the old floor is sound, you can proceed with your preparations for the new. Tap down and countersink any protruding nails. Secure loose boards by nailing them—directly into joists, where possible—with annular ring nails. Warped boards that can't be made level should be sanded flat or replaced.

If you're planning to install new wood block flooring (which is laid in adhesive), you're likely to have to remove all wax, varnish, paint, or other finish from the old floor. This may require a rough sanding with a commercial drum sander. (Floor sanders are available at most tool rental companies; see "Tools and Supplies," page 97.) For some types of new self-adhesive wood block, though, you may not have to remove an old finish, provided it's well bonded and hasn't started to flake. Check the manufacturer's recommendations for the particular kind of block you plan to install.

Old resilient flooring that is in reasonably good condition, flat, and securely fastened down can serve as a base for new wood block—but not for wood strips or planks. If the old floor covering is extremely worn or irregular, you'll be better off removing it before installing a new wood block floor.

To remove old resilient tiles, use a floor scraper. If the tiles don't come up easily, place a piece of paper over them and warm them with an old iron; heat will soften most adhesives.

To remove resilient sheet flooring, cut it into strips with a utility knife, taking care not to damage the subfloor. Peel or scrape away the strips, then scrape any excess backing off the subfloor with a wide putty knife or floor scraper. Fill low spots by troweling on a patching material.

Vinyl flooring laid with epoxy adhesives is extremely difficult to remove. It may be easier to cover the vinyl with ¼-inch underlayment grade plywood or hardboard. Fasten down the underlayment with 3-penny ring-shank nails or 4-penny cement-coated nails spaced 3 inches apart along the joints and 6 inches apart throughout the sheet. Stagger the joints and leave gaps about the thickness of a dime between panels.

Final preparations

New or old, the subfloor (or old floor, if that is to serve as the base for new flooring) should be cleaned thoroughly. Drive protruding nails below the surface with a nailset and hammer, and correct any other irregularities to make a perfectly flat base.

If you haven't already done so, remove the molding from around the baseboards, using a chisel and taking care not to split the wood (see illustration, page 49). As you remove molding and baseboards, number the pieces and the wall with chalk or a pencil so that you can replace them in their original positions.

If you've removed an old floor and plan to install block flooring that is thinner than the old flooring, lay an underlayment of untempered hardboard or plywood directly over the subfloor to raise the surface of the new flooring to the same height as the old. This will ensure a level transition between the new floor and adjacent rooms, and make it unnecessary to trim door casings, doors, or molding.

Finally, unless you're laying the flooring in mastic, cover the subfloor with a layer of 15-pound asphalt-saturated felt (butting seams) or soft resin paper (overlapping seams 4 inches). Trim the paper flush with the walls. As you put the paper in place, use a straightedge or a chalk line to mark the paper over the center of each joist. The markings will

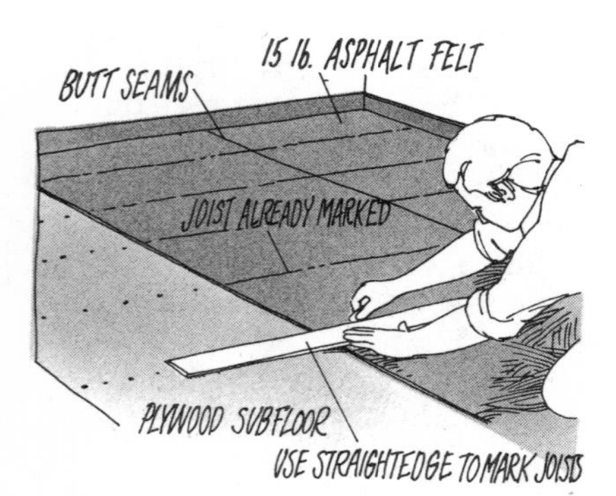

serve as reference points when you attach the new flooring. The paper will act as a moisture barrier, keep out dust, and help prevent squeaks in the new floor.

Protect the floor from heat. Flooring installed over the heating plant of a home needs extra protection to keep it from drying and cracking.

A *double* layer of 15-pound asphalt-saturated felt or a single layer of 30-pound felt will be effective. Or standard ½-inch nonflammable insulation board, cut to fit between joists from below, can be used as an alternative to heavier building paper.

Crawl space needs a barrier. If wood flooring is to be installed in a frame home with a crawl space, check the crawl space to see if a moisture barrier has been placed over the soil. If there is none, or if it has been torn or has deteriorated, use roofing material or 4 or 6-mil polyethylene sheeting to cover the ground (overlap edges by at least 2 inches). Use heavy objects such as stones or bricks to secure the moisture barrier in place.

Also check to see that the crawl space is well ventilated. There should be at least four vents, one in each corner of the building, with a total vent area of about 1½ percent of the floor area—more if required by code.

In dry climates where damp soil is not a problem, a moisture barrier isn't necessary, but ventilation is still important.

HOW TO INSTALL WOOD FLOORING

Once all the preliminaries have been taken care of and your new wood flooring has had time to adjust to its new home, you're ready to begin the actual work of installing the new floor. If you've prepared the subfloor properly and equipped yourself with the right tools and supplies, the job will go surprisingly fast. Below and on the following pages you'll find step-by-step instructions for installing a wood strip or plank floor. For directions on installing a wood block floor, see pages 53–56.

Installing strip and plank flooring

The procedures for laying tongue-and-groove strip flooring and tongue-and-groove plank flooring are almost identical. Both kinds are attached by blind nailing. With tongue-and-groove plank flooring, of course, screws and plugs can be added—partly for strength, but mostly for cosmetic reasons. Square-edged plank flooring, on the other hand, is secured solely with screws. Square-edged strip flooring is fastened with nails through the holes.

The best way to assure a tight floor is to install the strips perpendicular to the joists, which generally run across the room, and nail the strips through the subfloor into the joists. Use a chalk line to mark the locations of the joists on the paper covering the floor.

Plan the first course carefully. The key to a trouble-free installation of strip or plank flooring is to make sure the first course you lay will be parallel to the center line of the room.

Take several measurements of the width of the room and locate as accurately as possible the center line of the room. Snap a chalk line to mark the center—this will be your primary reference point.

Next, measuring from the center line near each end (be sure your measuring tape is placed perpendicular to the line), lay out and snap another chalk line about ½ inch from the wall you've selected as your jumping-off point (see illustration, top of next column). This line will enable you to align the first row of flooring correctly.

In rooms that are obviously irregular in shape, establish a

center line as close as possible to the center of the room and snap a chalk line. The first row of flooring will be laid from the center line of the room, with the grooves in the boards turned toward the center. A special wood strip called a spline (see illustration) is used to join two back-to-back grooved boards along the center line.

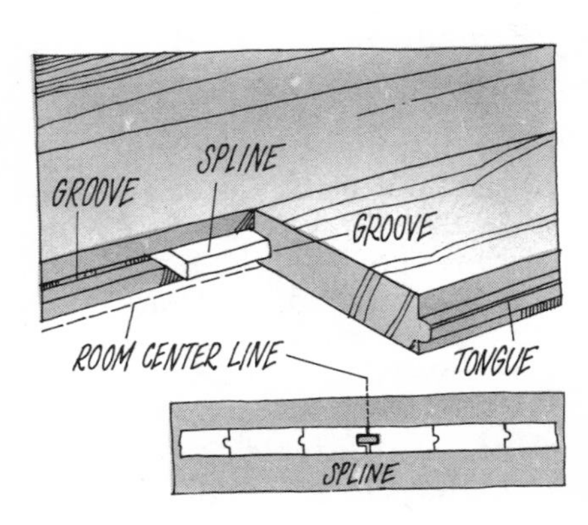

If you're working directly over screeds (see page 47), using a chalk line will not be as easy as it is with a solid subfloor, but the steps are the same. The screeds will be hidden, of course, by polyethylene film. In measuring to find the center and other lines (and at all times while you're laying flooring), avoid stepping between the screeds and puncturing the film.

Face nail the first row. Appearances notwithstanding, few rooms are perfectly square. As you'll be leaving a ½-inch space between the edge of the flooring and the wall, you can be compensated for some irregularities with no extra effort.

If you're starting from one wall of a room that's seriously out of square, though, it may be necessary to trim some individual boards so that your first row of flooring will line up properly while keeping the proper distance from the wall. If you're starting from the center in an irregular room, of course, the trimming will have to be done when you reach the walls.

Work with one bundle of flooring at a time. Lay the boards on the floor in the area you're working on, according to size. This will make it easier to plan the rows of flooring. Use the longest boards for the first row.

If you're starting along one wall, the first row of boards should be secured by face nailing where the nails will be covered with the shoe molding. Even

with square-edged planks, the first course should be tacked down with nails set close to the wall before you proceed with the business of screwing down the planks (see "Inserting screws and plugs," page 52).

Predrill the boards making holes slightly smaller than the diameter of the nails you'll be using, particularly if the boards are oak or another hard wood. This will make it easier to drive nails through them into the subfloor without splitting them.

If you're beginning at the center of an irregular-shaped room, you can start right off by blind nailing through the tongues of tongue-and-groove flooring (with a nailer, if you have one—see drawing, page 52) or screwing down square-edged planks (see "Inserting screws and plugs," page 52).

Plan several rows at a time. Once you've set your first row of boards, it will matter little whether you're nailing into a solid subfloor or into screeds. The technique is essentially the same.

Lay out boards six or seven rows ahead. This will make it simple to plan an effective and attractive pattern. End joints should be staggered so that no joint is closer than 6 inches to a joint in an adjoining row of boards. Find or cut pieces to fit at the end of each row, leaving approximately ½ inch between each end piece and the wall. As a general rule, no end piece should be shorter than 8 inches.

Longer boards might have bowed slightly because of moisture. With a simple block of wood and crowbar arrangement, these boards can be levered and held in place while you nail—or you

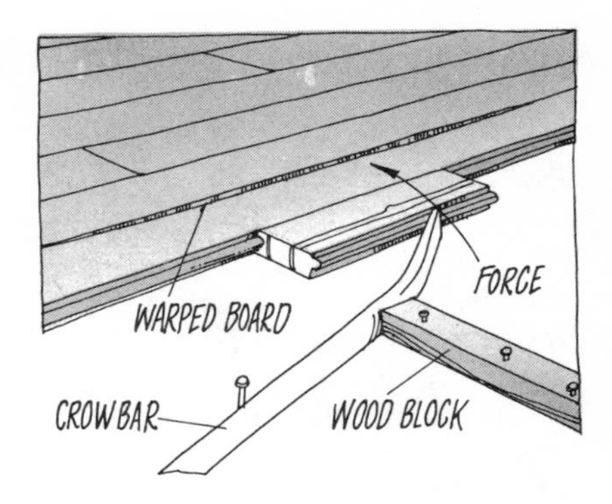

can use a device made of 2 by 4 scraps, two strap hinges, and two beverage can openers (see illustration).

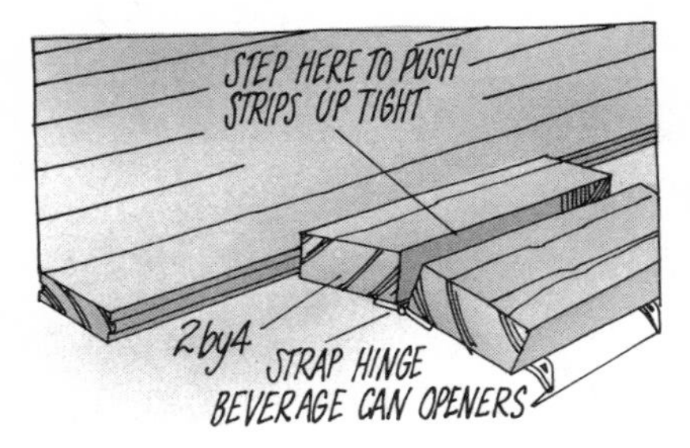

If you're attaching flooring to screeds, plan joints to fall over screeds whenever possible. Also, try to avoid having more than one joint in adjacent rows fall between screeds.

When laying flooring over plywood, avoid placing end joints in the flooring directly over joints in the subfloor.

Tighten rows as you go. As you place each row, take a block of wood, move it along the leading edge of the flooring you've just put down, and give it a sharp rap with a hammer before you drive a nail. Don't hit the block hard enough to damage the tongue—to be on the safe side, cut a groove in the block so that it will contact the flooring above and below the tongue (see illustration), or use a short length of flooring.

If you're installing planks, note that some producers of plank flooring recommend leaving a slight crack between boards—about the width of a putty knife blade. Follow the recommendations of your supplier for best results.

Periodically check the leading edge of the flooring as you work to make sure it's straight and still parallel to the center line.

Nail first few rows by hand. As you won't have enough space to use a nailer until you are several rows from the wall, you'll have to nail down the first courses by hand. (If you're installing plank flooring with screws and wood plugs, see the next section, "Inserting screws and plugs," for directions on attaching the boards to the subfloor.) By continuing to predrill holes for the nails, you can help yourself keep nails at the proper angle—45° to 50° to the floor—and also prevent splitting.

Take care not to crush the upper edge of boards by trying to drive nails flush with your hammer; these indentations will show when the boards have been joined. Instead, leave each nail head comfortably exposed; then place a nailset sideways over it along the upper edge of the tongue (see illustration), drive the nail home by tapping the nailset with your hammer, and, with the tip of the nailset, set the nail flush.

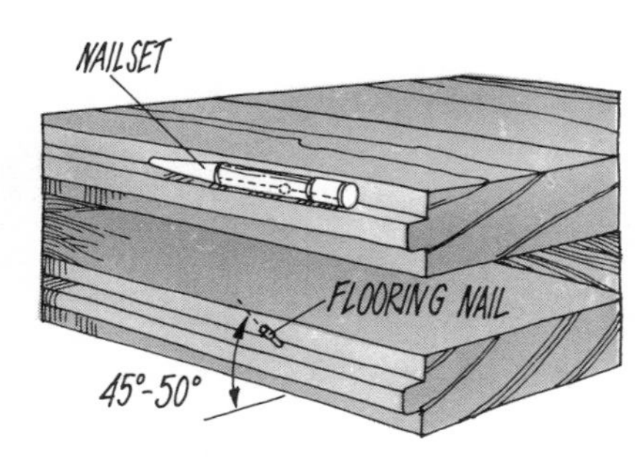

When nailing into screeds, drive nails into each screed along the full length of each board. Where boards are laid on top of overlapping screeds, drive nails into both screeds.

If you're nailing into a plywood subfloor, drive nails when possible directly through the plywood into the joists—especially if the subfloor is only ½ or ⅝-inch plywood.

Once you have laid and nailed the first few rows by hand, you can begin securing flooring with the nailer, which will automatically countersink all the nails it drives.

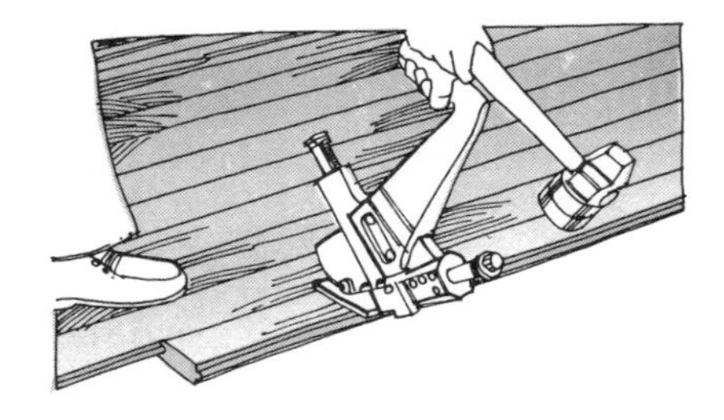

When you reach the last few rows, you'll find it difficult to toenail the boards. Predrill holes and face nail them.

Inserting screws and plugs in plank flooring is quite simple. With the plank in position, insert the screws in the predrilled holes and tighten; an electric drill with a screwdriver attachment will save wear and tear on your arm. If the boards aren't predrilled, mark the points where you intend to drill (see drawing for pattern), and use the center

punch to tap in starter holes. Using the electric drill with the power or brad point bit, drill each hole ¼ inch deep.

When the screws are in place, blow the dust out of each hole and fill the holes with wood plugs. Put a dab of common white glue on each plug before inserting it.

Plugs should either be set flush with the surface of the floor or left slightly protruding (they'll be sanded flat when the floor is finished), and held in place with a drop or two of white glue.

If you weren't supplied with plugs along with the flooring you bought, you can cut them from hardwood dowels of the same kind of wood as the plank flooring itself, or from a different kind of wood for contrast. For example, walnut plugs in an oak floor will create a striking effect.

Fitting in the last row. When you've progressed across the floor to the far wall (or from the center of an irregular room to either wall), the final strip of flooring must be placed so as to leave a ½-inch gap between the flooring and the wall.

If you're lucky, a standard board will fit. If not, you'll have to rip several boards down to the proper width.

Framing fireplaces and the like. Exposed obstacles such as fireplaces should be framed to give the job a more finished look. Cut pieces to fit, using a miter box to get accurate 45° angles (see illustration, top of next column). If you're working with tongue-and-groove flooring, plane off the tongue on those pieces that will run perpendicular to the flooring, then cut and place them so that the grooved edges are exposed. This will make it possible to insert the tongues at the ends of a few boards in the running floor into the grooves on the framing.

Floor openings covered with

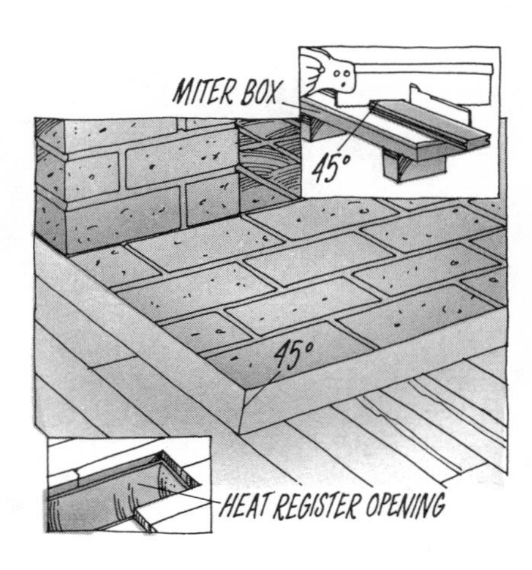

grates that have flanges need not be framed.

You may need a reducer strip. If your new floor will create a change of level from one room to the next, use a reducer strip for a smooth transition. A reducer strip (see illustration below) is milled with a rounded top. It will fit into the tongue of an adjacent

board, or, if laid perpendicular to the flooring pattern, can be butted against exposed board ends.

Finishing touches. If you've installed prefinished flooring, once the final board has been placed you can add (or replace) baseboards, shoe molding, and any grates that were removed. Molding should be installed with a slight gap between the flooring and the bottom of the molding; use a piece of thin cardboard for a spacer. Nail the molding to the baseboard, not to the floor, to let the floor expand or contract.

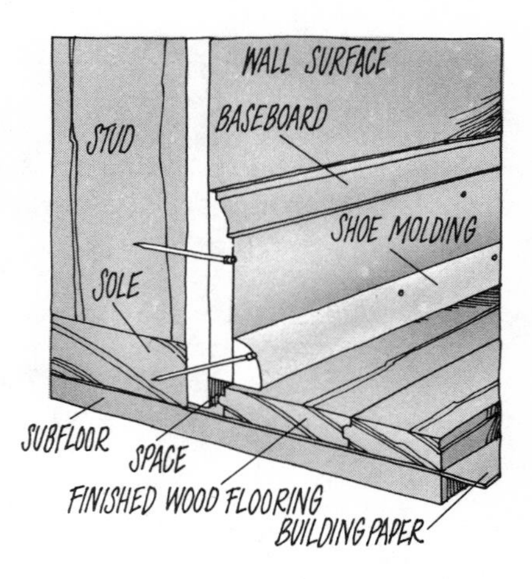

If you've installed unfinished flooring, the floor will have to be sanded and a finish applied before you replace the molding and so forth. These steps are covered in the section "Refinishing Wood Floors," beginning on page 96.

HOW TO INSTALL WOOD BLOCK FLOORING

In the next few pages, you'll find instructions for installing the basic types of wood block flooring. Because block flooring is produced in many different forms and can be finished in a variety of ways, it's important that you follow the manufacturer's recommendations.

A good wood block floor properly installed will have an exceptionally long life. Taking extra care to do a professional job of installation will pay off, giving you a handsome floor that will require little attention for many years.

Your first step, of course, is to prepare a proper base for your new wood block flooring. If you haven't already done so, read through "Preparing the Subfloor," beginning on page 46, for instructions on preparing various types of subfloors for wood block.

(Continued on page 55)

HOW TO ESTABLISH AND USE WORKING LINES

The key to installing flooring that comes in squares—resilient tile, wood block, ceramic tile, and some masonry units—is first to establish accurate working lines.

With flooring that is set in adhesive, mastic, or a thin bed of mortar, you'll start your installation at the center of a room and work toward the walls. The instructions that follow will show you how to lay out and use working lines for this type of installation. Flooring units set in a traditional thick bed of mortar are laid differently—you begin at one wall and work across the room; for instructions on establishing and following working lines for this kind of installation, see "Starting at the wall," page 69.

You can't just lay out working lines parallel to the walls of a room; although the wall may appear to be parallel and meet at right angles, few rooms are perfectly rectangular.

Begin by locating the center point on each of two opposite walls, and snap a chalk line across the floor between the two points. Then find the centers of the other two walls and stretch your chalk line at right angles to the first line, but snap the line only after you've used your carpenter's square to determine that the two lines cross at precise right angles.

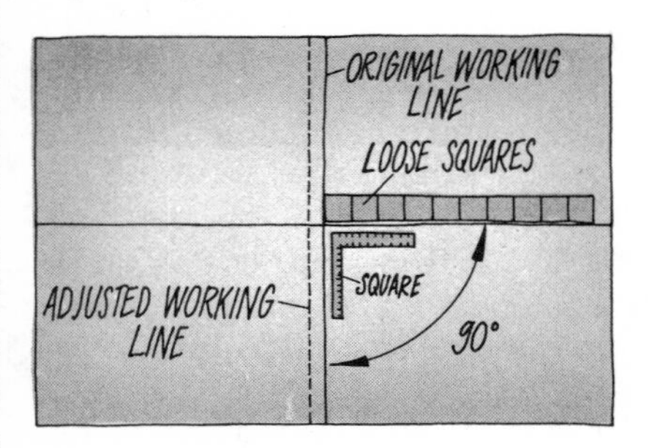

Next, to ensure that the border units around the perimeter of the room will have a balanced look, lay a row of loose squares of flooring along each line from wall to wall. Allow for space between squares if you're installing ceramic tile or masonry units that require grouting or mortar joints. If the space between the last square and the wall is less than half a unit wide, move the center lines the width of a half-tile. Check by laying squares loosely on across to the opposite wall to assure you have adequate spacing.

If you prefer to lay your flooring in a diagonal pattern, establish working lines as described above. Then mark each line at two points 4 feet from the center on either side. From these points, measure and mark the end of a 4-foot line in each direction, taking care to lay your tape measure exactly perpendicular to the center lines.

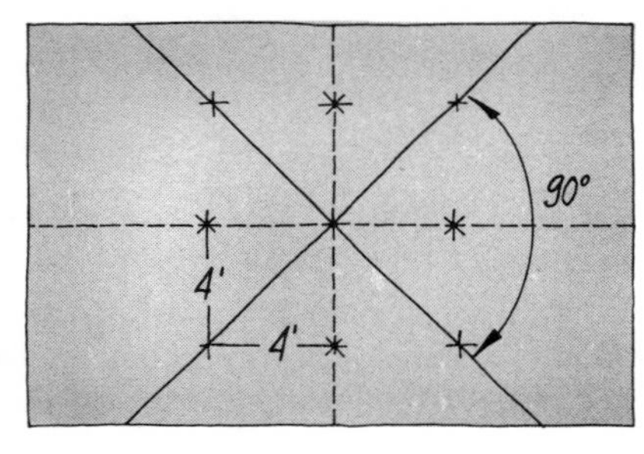

You should now have an 8-foot square, the center of which corresponds with the center of the room. Snap chalk lines between opposite corners of the square to get diagonal working lines. If your measurements were accurate, the diagonal lines will intersect exactly at a right angle over the center of the room. If they don't, recheck your measurements.

As a final test to make sure the lines are perpendicular, whether they're running parallel with the walls or on a diagonal, measure 3 feet along one line and 4 feet along the other, then measure diagonally between these two points; this distance should be exactly 5 feet.

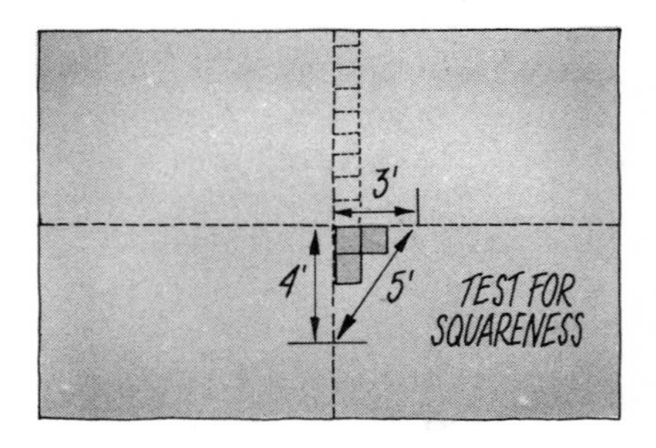

Regardless of the material being installed, the sequence for placing squares is the same. If you're using a square pattern (see illustration), you can fill in the floor area in quarters or lay flooring over half the room at a time.

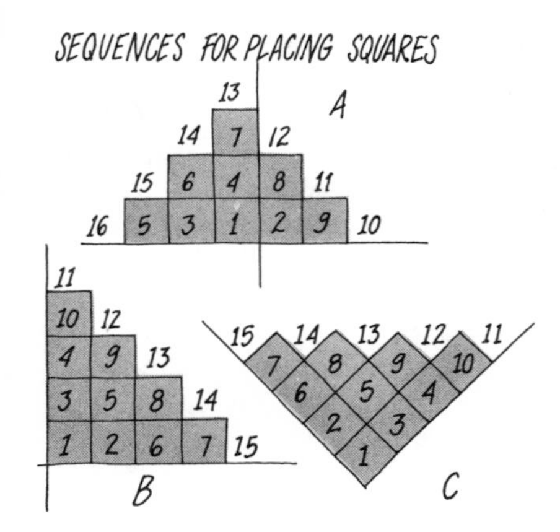

...Continued from page 53

The installation details that follow cover the essential steps common to most block flooring projects.

Plan working lines. No matter how they may look to the naked eye, few rooms are perfectly symmetrical. All flooring suppliers recommend that you establish working lines independently of the wall lines.

As most flooring manufactured in squares—wood block, ceramic tile, and resilient tile—is laid out in much the same way, working from the center of the room, refer to "How to Establish and Use Working Lines," facing page.

Spreading the adhesive. In planning your wood block installation, take note of the "open time" of the adhesive you will use—the time you'll have between spreading the adhesive and placing the wood blocks.

When you're ready to start spreading the adhesive, use a putty knife to transfer some of the adhesive from the container to the spot on the floor where you plan to lay the first block. Holding the notched trowel at a 45° angle, spread the adhesive evenly, putting firm pressure on the trowel. You should be able to see your working lines between ribbons of adhesive; if you can't, take care to spread the adhesive up to, but not over, the chalk lines.

How much adhesive you should spread at any one time will depend on the open time of the adhesive and the pattern you choose to follow in laying the block (again, see "How to Establish and Use Working Lines," facing page).

Placing block flooring. When laying block flooring, take care to align each block carefully. Minor irregularities can quickly become major problems as you work your way across the floor.

The illustrations that follow use placement sequence A illustrated on the facing page. You can adapt these instructions to the other two sequences shown or one suggested by the manufacturer. A step pattern allows you to lay each tile in a corner formed by two others, making it easier to maintain straight lines.

Place the first block at the intersection of the two working lines, aligning the edges (grooved edges if tongue-and-groove block) with the lines. Take

special care in placing the first ten blocks; their alignment will determine the appearance of the entire finished floor.

Set the second block against the first, as illustrated. Align it carefully with the first tile and your working lines. If you're working with flooring that has tongues and grooves, engage them as you put the blocks together—*don't slide the blocks into place*. Sliding the blocks will force adhesive into the joints, making it difficult to keep the blocks properly aligned.

Continue placing the blocks in the pattern as shown in sequence A. After you've set several blocks, tap them sharply in several places with a rubber mallet to seat them firmly in the adhesive. If any adhesive is forced up between the blocks or tracked onto the floor surface, clean it off immediately with thinner.

By the time you've laid several rows of block, it may be necessary to walk or kneel on the newly laid flooring. To keep the blocks from sliding out of position and to prevent adhesive from being forced up between blocks, place pieces of plywood over the new blocks to distribute your weight more evenly.

Trimming blocks to fit walls and obstacles. When you reach a wall or other obstacle, it will probably be necessary to cut individual blocks to fit. Allow for the natural expansion of the wood by leaving a ½-inch gap between the block and the wall; with other obstacles, leave a gap of about 1/16 inch. Some flooring manufacturers supply cork expansion strips to fill the space between the block and the wall; use this material if it's available.

To cut a block to fit, mark it as shown below; a piece of scrap

wood of the right thickness can be used as a spacer. Then use a crosscut or saber saw to trim the block. Cut and set the blocks one at a time as you proceed.

Where it's necessary to cut off

the bottom of the casing around a doorway, use a block for a guide to make the proper cut. Slide the

block under the casing to install it.

Where new flooring meets old. If your new flooring will be level with the floor in an adjacent room, the new and old floors should meet under the door. The joint can be hidden with a wood saddle, available from your flooring supplier.

If the new flooring is higher than the floor in an adjacent room, install a reducer strip (see illustration, page 53) to adjust the level where the floors meet. This too can be supplied by the flooring manufacturer in the same wood and finish as the block.

Finishing the job. Allow newly installed wood blocks to set overnight—or as long as recommended by the manufacturer. Then replace any baseboards, shoe molding, and grates that were removed. Leave a gap between the moldings and the floor (use a thin piece of cardboard as a spacer). Nail the molding to the baseboards, not to the floor. This will allow for the natural expansion of the flooring. If the floor level has been raised by the new flooring, it may be necessary to trim material off the bottoms of doors before you rehang them.

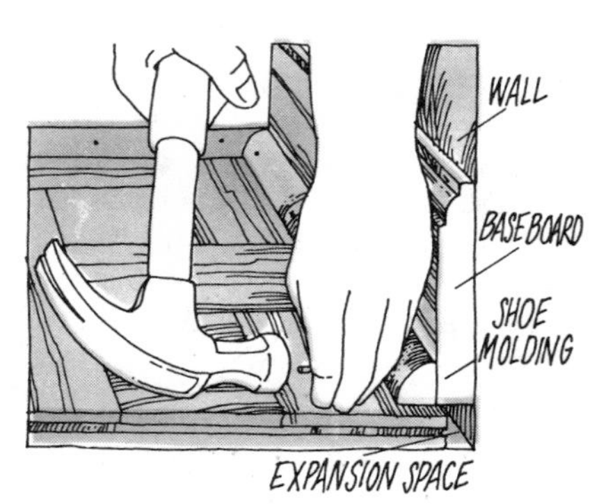

CARING FOR A WOOD FLOOR

Water is the natural enemy of wood. Since water seeping between boards can cause stains or swelling, avoid wet-mopping wood flooring and using water-base waxes.

In general, vacuum or dry-mop wood flooring about once a week, or as often as you vacuum carpeting. Most wood floors should be waxed once or twice a year.

Wood flooring is finished with either a penetrating sealer or a surface finish.

If you've installed a floor with a factory-applied finish, the manufacturer will specify whether the wood has been finished with a penetrating sealer or a surface finish. But if you don't know how your floor was finished, assume a surface finish was used. Treating a penetrating sealer as a surface finish can do no harm.

Cleaning floors finished with penetrating sealers. For occasional cleaning up of spots, use mineral spirits and steel wool. After wiping the area with a clean, soft rag, buff the floor lightly.

When the floor shows signs of excessive wear, clean a small area at a time with a mineral-base cleaner, going over the area with rags or #3 steel wool. Wipe off any excess cleaner with a clean rag. After a few hours, buff the floor with a bristle buffing pad.

Finish with a good paste wax (one intended for use on hardwood floors) and buff the floor.

Cleaning surface-finished floors. For periodic cleaning of spots and stains, go over the floor with a barely damp sponge mop. If any stains remain, use a mild cleanser, such as ammonia or white vinegar, to remove them. Rinse the floor with a clean, damp (not wet) sponge mop to remove any residue. To restore the floor's luster, you can buff it lightly.

For a more thorough cleaning, use a mineral-base cleaner in the same way as described for floors finished with penetrating sealers.

Though manufacturers of polyurethane finishes claim no waxing is necessary, floor care experts agree you'll get better wear and appearance if you wax the surface once or twice a year.

Apply a light coat of paste wax (suitable for use on hardwood floors) and buff the surface to a luster using a buffing machine or clean, soft rags.

RESILIENT—EASY TO WORK WITH

For the homeowner who hasn't bought new flooring for several years, looking at what's available today in resilient materials will provide many pleasant surprises. Flooring manufacturers have made major strides in developing new products that are attractive, long-wearing, and easy to install.

This section includes a discussion of the two basic types of resilient flooring available, instructions on how to order each type, a description of the tools and supplies you'll need for installation, and step-by-step instructions for preparing the subfloor and installing resilient flooring.

THE TWO TYPES: SHEET AND TILE

Resilient flooring comes in two forms—in sheets up to 12 feet wide and in standard 12-inch squares.

Sheet flooring can be laid in adhesive or loosely, like a rug. Though a few types of flooring are available in widths up to 12 feet, most sheet flooring is 6 feet wide so you may have to make seams. Individual tiles are either laid in adhesive or purchased with self-sticking backing that makes adhesive unnecessary.

Both sheet and tile resilient flooring are available with smooth or textured surfaces in plain colors or in patterns. The hundreds of patterns you can buy include authentic-looking imitations of all types of flooring—brick, slate, wood, marble, terrazzo, flagstone, and ceramic tile. Because there are so many choices, visit your local flooring materials dealer and take the time to explore all the possibilities carefully.

Ease of installation should be a major consideration in your choice of flooring. Most manufacturers will provide specific instructions for installing their products and will specify the kind of adhesive to use. Look over these instructions before making a final choice.

Advantages of sheet flooring

Resilient sheet flooring can normally be installed with few seams—its greatest advantage over individual tiles. This makes sheet flooring particularly practical for bathrooms, kitchens, laundry rooms, and entryways—rooms where water can be spilled or tracked onto the floor. Even though modern adhesives make it possible to lay a perfectly bonded tile floor, water can work its way down through the smallest gaps between tiles that haven't been properly set.

Working with a large single sheet of flooring is more difficult than working with individual tiles. But modern resilient sheet flooring, particularly the foam-backed and the thin vinyl types, is considerably easier to handle than the stiff, brittle floor coverings that were commonly used a few years ago. The properties of various types of sheet floor coverings are discussed in detail in the introduction to this book (see page 6).

All resilient sheet floor coverings are installed in much the same way. The primary variable is whether or not an adhesive is used. All major suppliers of resilient sheet flooring have developed products that can be laid without adhesives.

Tile flooring

When it was first introduced, asphalt floor tile was very popular for homes because it was inexpensive and came in easy-to-handle squares. But it was hard to keep clean and often cracked or chipped.

Resilient tiles made of vinyl-asbestos, solid vinyl, or rubber have made asphalt tile almost obsolete. Asphalt tile is still used in seldom-used rooms, but most builders and homeowners have turned to newer materials that are easier to maintain and last longer.

Tiles are usually set in adhesive, but they can be bought with self-sticking backing. This adds slightly to the cost, but sparing themselves the sometimes messy business of applying adhesive is worth the extra investment to many homeowners.

ORDERING RESILIENT FLOORING

If you're buying sheet flooring, your dealer will, of course, need to know at least the rough dimensions of the floor you plan to cover. But if you provide a fairly detailed map of the room, the dealer may be able to give you a head-start on your installation by making the first rough cuts on your sheet flooring.

If you're planning to install sheet flooring in a large room, you may have to seam two sheets together. Your dealer will be able to help you decide how much flooring you'll need, how best to arrange adjoining sheets to cover the floor with the minimum amount of material, and how to ensure that the pattern in adjoining sheets will match. Read "Cutting and installing flooring with seams," page 61, before ordering so you'll know what's involved.

Resilient tiles come in a standard 12-inch square size except for asphalt tiles which are available in 9-inch squares. Once you choose the tile you want, your flooring materials dealer will be able to tell you exactly how many you'll need to cover your floor if you provide the basic dimensions.

Be sure to allow for waste. If you're doing a large floor—over 1,000 square feet—you'll want a spoilage margin of about 3 percent. If you're covering a small floor that's between 50 and 100 square feet, you'll want to order 10 to 12 percent extra. In any case, it's a good idea to put aside a few pieces in case it's ever necessary to replace damaged tile.

When you order, ask your dealer what the policy is for returning unused tiles. Unless you've ordered tile not normally carried by your supplier, full cartons are usually accepted for credit. Some dealers will accept returns of odd amounts if the tile is one of their popular styles.

Whether you're buying sheet or tile flooring, plan to have it delivered at least two or three days before installation. Resilient flooring, especially the sheet type, will expand and contract as the temperature rises and falls, so new flooring should be stored in a warm, dry place—ideally, in the room where it's to be installed.

TOOLS AND SUPPLIES

To install resilient sheet or tile flooring you'll need a steel tape measure or folding rule, a utility knife (a special hooked blade for cutting sheet flooring is available) or linoleum knife, a putty knife, a pair of heavy-duty scissors or shears, a steel straightedge, clean rags, and—if you're going to use it—adhesive and the solvent recommended by the adhesive manufacturer. You'll also need a special notched trowel to spread the adhesive; your flooring dealer can recommend the trowel you should use.

If you're putting sheet flooring in a large room, you may need a seam kit to join adjacent pieces of flooring. If you're installing individual tiles, you'll need a chalk line and square to establish working lines on the subfloor. Consider renting a floor roller from the flooring dealer or from a tool rental company if you're installing new sheet flooring in a larger than average-size room.

PREPARING THE SUBFLOOR

Since most resilient flooring is so pliable, it will conform to irregularities in the subfloor. For this reason, it's essential that care be taken to prepare the subfloor properly to guarantee a smooth surface. It's also important to protect the subfloor, especially if it's concrete, against moisture. Resilient flooring can be laid on concrete slabs (on or below grade), on wood subfloors made of plywood panels or individual boards, or directly on an old floor.

Before beginning the preliminary work of preparing the subfloor for new flooring, check the basic floor and supporting structure to make sure it's in good condition. For tips on what to look for, see the sections on problems and repairs beginning on page 87. Read these sections carefully and make the series of checks recommended. It's important to discover and repair minor problems now before they become more serious.

If the basic floor structure appears to be in good shape, you can go ahead with your flooring project confident that the new floor will have a long life.

Concrete slabs must be dry, clean, and level

If you're laying resilient flooring over a concrete slab (on or below grade), you'll have to make sure the slab is level and free from dirt, grease, old finishes, and any other foreign matter. In addition, check that the slab is completely dry and take steps to ensure it will stay dry. Any moisture coming through the concrete even after the flooring is laid will eventually cause the flooring to loosen.

Most newly poured concrete floors—even those poured over a moisture barrier—need at least a month to dry. Sometimes, drying can be accelerated with proper heat and ventilation. But because concrete has been known to retain moisture for as long as two years, it's wise to test for moisture before proceeding with your flooring installation (see page 46).

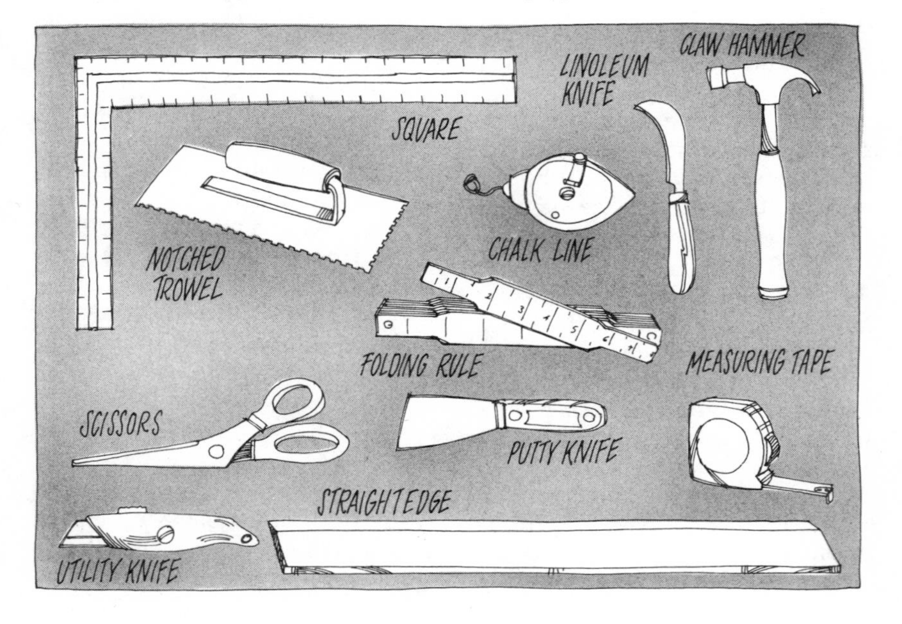

Once you're satisfied that the concrete slab is completely dry, check it for any grease, oil, old paint, or other materials. To remove grease and oil spots, use a chemical garage floor cleaner commonly sold at most auto supply stores. Use a putty knife or other flat-bladed tool to chip away any blobs of plaster or other hardened materials you may find on the surface of the slab.

To remove old paint or old sealers, sand the floor to bare concrete, using a floor sander equipped with #4 or #5 open-cut sandpaper. (For tips on renting and operating a floor sander, see pages 97 and 99.) Vacuum or sweep the floor as you proceed.

Once the slab is as clean as possible, use a straightedge to find any low spots in the surface. Fill these uneven areas and any cracks or joints with a latex underlayment compound; follow the manufacturer's directions for its use. When the compound is dry, brush the patched areas with a dry, stiff bristle or wire brush and sweep or vacuum up all loose material.

If the slab is too uneven to be completely leveled by patching, you'll have to resort to one of two alternatives. The first is to lay a new plywood subfloor on screeds (see "How to prepare a base for wood flooring over a slab," page 47). The second is to pour a new, thin concrete slab over the old. For details on what is involved in pouring a concrete slab, see the *Sunset* book *Basic Masonry Illustrated*.

As a final step in preparing a concrete subfloor, cover the surface with a sealer or other moisture barrier; your flooring dealer can suggest the best one for your situation.

Preparing a wood subfloor

Wood subfloors are made from plywood panels or individual boards. The two types require similar preparations before being covered with resilient flooring. With either, be sure to clean the surface carefully before applying any floor sealer.

To avoid the damaging effects of moisture, the wood subfloor over which the flooring is laid must be adequately suspended and well cross-ventilated from underneath.

Preparing a plywood subfloor for new resilient flooring is relatively simple. If the subfloor is new, make sure the panels are securely attached with annular ring nails or cement-coated nails, and that the nail heads are flush with the surface. To fill minor indentations and gaps larger than the thickness of a dime between panels, use wood putty; allow the putty to dry thoroughly before sanding smooth.

If you have to take up old flooring to make way for the new floor, you must remove every bit of felt backing, grout, or any other material that might be sticking to the subfloor. Most of this material can be scraped loose with a wide-bladed putty knife. But you may need a solvent to remove stubborn particles of felt backing. Soften old adhesive by heating it with an iron (to prevent the adhesive from sticking to the iron, cover the area with a piece of paper first).

If you find that the old covering is particularly difficult to remove, consider covering the floor with a new underlayment of plywood or untempered hardboard at least ¼ inch thick (see below). In selecting new underlayment, take into account the level of adjoining floors.

Preparing a wood board subfloor. Because most wood board subfloors are extremely difficult (if not impossible) to make smooth and level enough for a resilient floor covering, they should be covered with ¼-inch underlayment-grade plywood or untempered hardboard.

Before you begin, check the subfloor carefully and make any necessary repairs. Renail any loose boards to make the subfloor as level as possible. If boards are badly cupped or bowed, you might have to replace them. Boards that aren't so misshapen can often be sanded smooth (details on renting and operating a floor sander are on page 97 and 99). Minor indentations in the subfloor need not be repaired because the plywood or hardboard underlayment will cover them.

Whether plywood or hardboard, underlayment should be installed in 4 by 4 or 4 by 8-foot sheets—whichever are easier to handle. As you install the panels, keep these three points in mind:

- Stagger joints between panels to avoid having four corners meet (see illustration, page 48).
- Leave gaps about the thickness of a dime between panels to allow for expansion of the material.
- Allow ⅛-inch clearance between panels and walls or baseboards.

Fasten the panels down with 3-penny ring-shank or 4-penny cement-coated nails spaced 3 inches apart along the edges and 6 inches apart across the face of each panel. Always start nailing in the center of each panel and work out toward the edges.

Using an old floor as a subfloor

You can put new resilient flooring over an old floor that's been properly prepared to provide a clean, level base.

Whatever kind of old floor you're dealing with, you'll first have to move out all the furniture and remove register or vent covers and the shoe moldings (perhaps the baseboards, too—see "Preparing the room," next page).

...Continued from page 59

Preparing old resilient flooring. Resilient flooring can be installed directly over old resilient flooring only if the old floor is completely smooth, solid (not cushioned), and tightly bonded to the subfloor.

If there are no signs of damage caused by moisture and if surface damage is limited to a few loose tiles or small areas where sheet flooring has worked loose, then repairing these minor problems (see page 107) and covering the old flooring may be the answer.

Before installing new flooring over old, always clean the surface of the old floor thoroughly, removing any old wax or other finish. *Caution:* Do not attempt to sand an old floor covering as a short cut to removing old finishes; certain types of resilient flooring contain asbestos fibers that can be extremely harmful if inhaled.

Preparing old wood flooring. Resilient flooring can be laid directly over a hardwood floor if it's level and in good condition. Uneven wood floors may have to be rough-sanded with a floor sander to provide a suitable base for new flooring.

If the old wood floor is in poor condition, you should install a covering of plywood or hardboard underlayment. Follow the instructions given on the previous page for covering a wood board subfloor.

Preparing other types of old flooring. Resilient flooring should never be placed directly over old ceramic tile, slate, or masonry flooring with an uneven surface. These kinds of old flooring should be removed, if possible, unless they're reasonably level; then the old floor can be covered with a plywood subfloor laid on screeds (see page 47). Solid masonry-type floors can be covered with a new concrete slab.

INSTALLING RESILIENT FLOORING

On the next few pages you'll find step-by-step instructions for laying both sheet and tile resilient flooring.

Checking tools and supplies. Before beginning any work, make sure you have all the tools and supplies you'll need for the job (see page 58). If you're planning to install your flooring with adhesive, carefully read the adhesive manufacturer's instructions and make sure you have the correct solvent on hand for cleaning up spills and smudges.

Preparing the room. Remove all furnishings from the room. If you haven't already taken up shoe molding, vent or register covers, and other fixtures, remove them now. Baseboards should be removed only if they can be loosened without damaging the walls or doorjambs. If the baseboards are faced with shoe molding, only these need to be removed, unless you're replacing the baseboards with a vinyl wall base (see page 62). If you take time to label or number the pieces of molding you remove, you'll have an easier time replacing them later.

You should also decide now between the two options for trimming your new flooring around the room's door casings. As discussed under "Trimming around doorways," page 62, you can either cut away just enough of the bottom of the door casings to slip the flooring underneath (the easiest method), or trim the flooring itself to fit flush with the casings. If you want to go the first route, trim the bottoms of the door casings now.

Laying sheet flooring

Installing resilient sheet flooring is as much a matter of deliberate planning and careful trim work as of hard labor. Whether you're using adhesive or laying the flooring loose, the procedures are the same.

You'll need to begin by drawing an accurate floor plan of the room (see below). Using this plan, either you or your flooring dealer will make the first rough cuts. Instructions are given on the next page for installing a single piece of flooring and for installing flooring with seams.

Laying flooring without applying adhesive to the subfloor is easy. You will need to make a plan and cut the flooring (see below). Then you can simply roll out the flooring and shift it until it has been moved into the proper position. Some manufacturers recommend using adhesive around the edges or stapling the edges in place.

Making a plan. To ensure a clean, professional installation, it's essential that you take time to make an accurate floor plan and map out the best approach to installing the flooring, especially if you need to use more than a single sheet of flooring.

Take a piece of graph paper and make a scale drawing of the room, using exact measurements. Include the locations of doorways, alcoves, closets, counters, and any other irregularities in the room. Having a partner will help make the job of measuring and mapping go more quickly. If your room is very irregular, you may want to make a full-size pattern of the floor instead of a scale drawing.

Though resilient sheet flooring can be purchased in widths up to 12 feet, it may be necessary to make a seam between two pieces to cover a large area. Looking at your floor plan, determine how to combine sheets so you can cover the floor with the minimum amount of material. If the flooring is patterned, you'll need enough to match the pattern at the seams.

Cutting and installing flooring without seams. The most critical step in laying sheet flooring is making the first rough cuts accurately.

Unroll the flooring in a large room or in a clean garage or basement. Transfer the floor plan directly onto the top of the new flooring, using chalk or a water-soluble felt-tip pen, a carpenter's square, and a long straightedge.

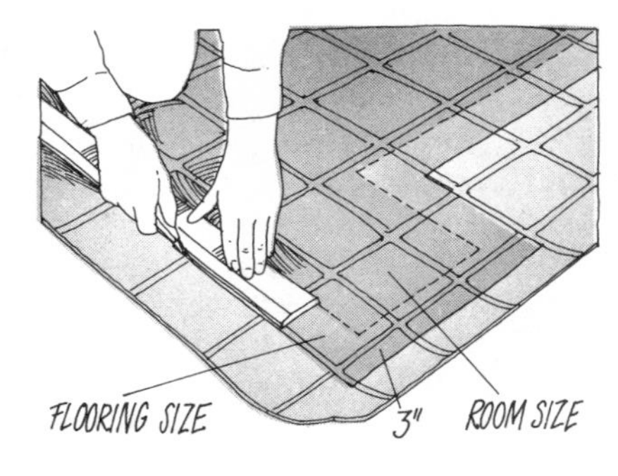

Using a linoleum or utility knife and heavy-duty scissors or shears, cut the flooring so it's roughly 3 inches oversize on all sides (the excess will be trimmed away after the flooring has been put in place).

If you're laying sheet flooring that must be secured with adhesive, apply the adhesive to the subfloor, following the directions of the adhesive manufacturer. Adhesive can either be spread over the entire subfloor at once, or, depending on the type of adhesive used, spread in steps as the flooring is unrolled. Check the adhesive's open time—the time it takes for the adhesive to dry.

Carry the roll of flooring into the room and lay the longest edge against the longest wall, allowing the 3-inch excess to curl up the wall. Check to see that the flooring curls 3 inches up each adjoining wall. If the entire floor has been covered with adhesive, slowly roll the flooring out across the floor, taking care to set the flooring firmly into the adhesive as you go. Or you can work your way across the floor spreading adhesive and unrolling the flooring as you go.

Cutting and installing flooring with seams. The flooring dealer can make the first rough cuts for you if you supply the floor plan.

To cut the flooring to size yourself, transfer the floor plan you've made to the flooring material; use chalk or a water-soluble felt-tip pen, a square, and a long straightedge. On patterned flooring, be sure to leave the margins necessary to match the pattern on adjoining sheets at the seam. If your flooring has a simulated grout or mortar joint or any other straight line, plan to cut the seam along the midpoint of the printed joint. When you order your flooring, your supplier will be able to tell you if the design you've selected requires reversing the sheets at the seam.

Using a linoleum or utility knife and heavy-duty scissors or shears, cut the piece that requires the most intricate fitting first; cut the piece so it's roughly 3 inches oversize on all sides. If adhesive is being used, spread the adhesive on the subfloor as directed, but stop 8 or 9 inches from the seam. Then position the sheet on the floor. If no adhesive is being used, simply put the first sheet in place.

Next, cut the second sheet of flooring and position it carefully so it overlaps the first sheet at least 2 inches; make sure the design is perfectly aligned.

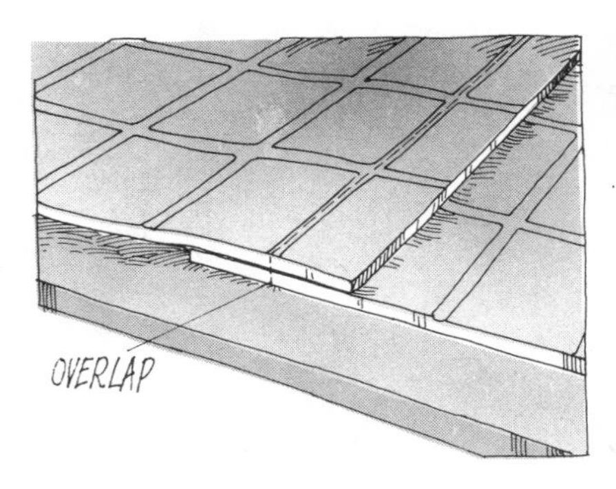

If adhesive is being used, roll up the flooring and spread the adhesive over the remainder of the floor, stopping a few inches from the edge of the first sheet of flooring. Reposition the second sheet of flooring, starting at the seam overlapping the edge of the first sheet; again, take care to align the design perfectly. Then roll the flooring out, setting it into the adhesive.

If you're not using adhesive, position the second sheet carefully; then lift it up and secure it to the subfloor with two or three strips of double-faced tape.

When the flooring is in position, trim away excess flooring at each end of the seam in a half-moon shape so the ends butt against the wall (see illustration).

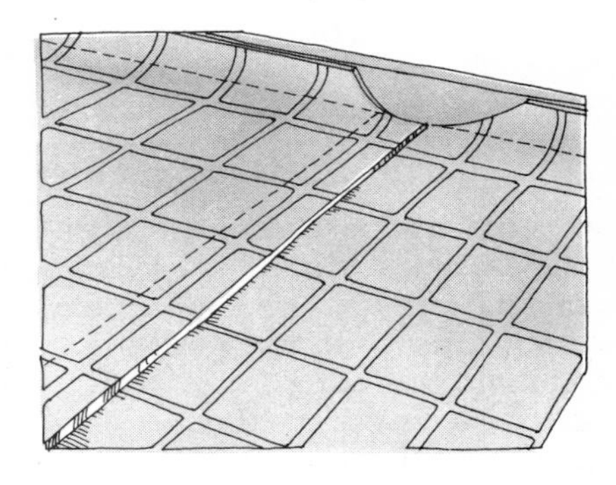

Using a steel straightedge and a sharp utility knife, make a straight cut (about ½ to ⅝ inch from the edge of the top sheet) down through both sheets of

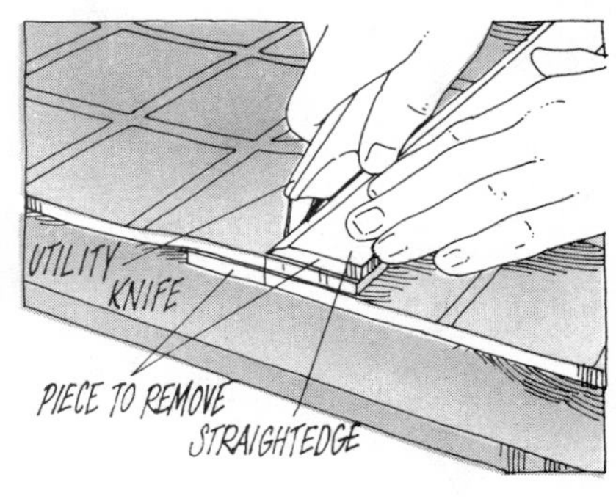

flooring. Lift up the flooring and spread adhesive on the subfloor under the seam.

If no adhesive is being used,

apply a long piece of double-faced tape under the seam. Roll the flooring away from the seam; then, removing one side of the backing paper from the tape as you go, stick the tape to the subfloor, stopping 5 inches from each wall. Peel off the remaining backing paper and press the flooring down firmly over the tape, keeping the edges close together.

Clean the area around the seam—use solvent if adhesive has been squeezed up between the pieces of flooring. When the seam is dry and dirt free, use the seam sealer recommended by the flooring supplier for your specific type of flooring to fuse the two pieces.

Trimming to fit corners and walls. When the flooring has been positioned, you'll need to make a series of relief cuts at all outside and inside corners so the flooring will lie flat on the floor.

At outside corners, start at the top of the lapped-up flooring and cut straight down to the point where the wall and floor meet. At inside corners, cut the excess flooring away with diagonal cuts, gradually trimming it away until the flooring lies flat in the corner (see illustration).

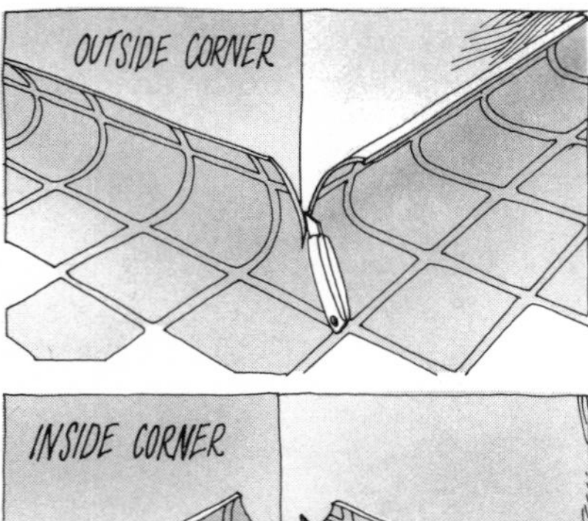

Once the corners have been trimmed, the material lapped up against the walls must be removed. Using an 18 to 24-inch-long piece of 2 by 4, press the flooring into a right angle where the floor and wall join.

Lay a heavy metal straightedge along the wall and trim the flooring with a utility knife, leaving a gap of about ⅛ inch between the edge of the flooring and the wall to allow the material to expand without buckling. If you're planning to attach a wall base, be sure the base will overlap the edge of the flooring by at least ¼ inch.

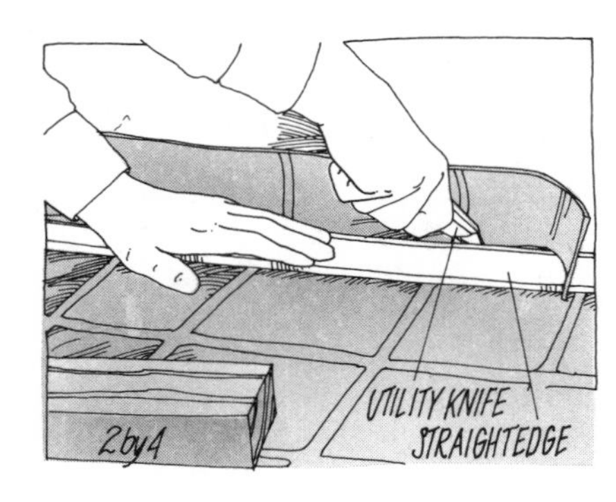

Trimming around doorways. The flooring should fit snugly around doorways and any other obstacles where there will be no floor molding to provide a finished appearance. Around a doorjamb, the most effective way to hide an exposed edge is to cut away just enough of the door casings to slide the flooring under the casings (see illustration).

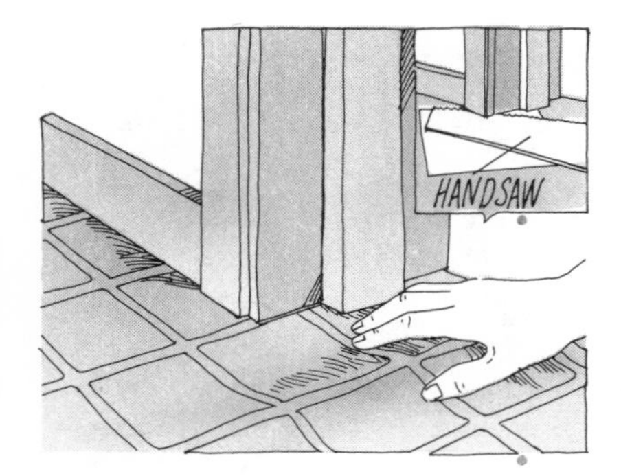

Finishing the job. When the flooring is in place, clean the surface carefully. The manufacturer may recommend a specific floor cleaner that won't harm the flooring you've installed. Use a recommended solvent to wipe away any adhesive still remaining on the surface.

If the flooring has been laid in adhesive, make sure the flooring is set firmly into the adhesive. Start at the center of the room and work out any air bubbles that may remain. A floor roller can be used effectively for this purpose.

When the floor is clean, flat, and well settled, replace any baseboards that have been removed. Then reattach the shoe molding, leaving a 1/32 to 1/16-inch gap between the floor and the bottom of the molding. Always drive nails through the molding into the baseboards, not down through the flooring, so the flooring can move slightly without buckling. Replace any vent covers or fixtures that were removed.

An alternative to baseboards and shoe molding is vinyl wall base. Available in a variety of complementary colors, wall base is installed using the same tools required to install resilient flooring, but special adhesives made for use on vertical surfaces are necessary to attach the base.

Vinyl wall base is flexible and usually comes in rolls. It should be fastened directly to the wall with the lower edge of the base resting on the resilient flooring. It should *not* be attached to the surface of the floor.

To install, spread the adhesive with a notched trowel or a brush according to the recommendations of the manufacturer. Allow the adhesive to set as directed; then, starting at an inside corner, begin installing the base by simply pressing it firmly in place until it holds.

At outside corners, stretch

the vinyl around the corner, pressing it firmly against the wall. To form inside corners, cut the vinyl to fit; use a sharp utility knife to miter the ends. To form an inside corner without a joint, score the back of the vinyl and bend it to fit snugly into the corner.

Finish the sheet flooring as recommended by the manufacturer.

Laying individual tiles

Installing resilient floor tile is one of the more common do-it-yourself projects tackled by the average homeowner. By following the directions on the next few pages, you can do a professional job of creating a new floor with tiles.

Establishing working lines. When you're satisfied the subfloor is sound and properly prepared and you've made other necessary preparations in the room, you can establish your working lines—the guides you'll need to place your tile accurately (see "How to Establish and Use Working Lines," page 54).

Working with adhesives. If your tiles are self-sticking, you're ready to place them (see next column). If you're going to be using adhesive, check the instructions on the container for the recommended method of applying the product. Since adhesive is most commonly spread with a notched trowel, the instructions below follow that method. If you're working with an adhesive that's spread with a brush or roller, follow the manufacturer's directions.

Be sure to consider the adhesive's open time—the amount of time you'll have to set the tile before the adhesive dries. The adhesive's open time will determine how large an area of the floor you should cover with adhesive at one time.

Using a putty knife or broad spackle knife, begin spreading adhesive by transferring a small amount to the spot on the floor where you plan to lay the first tile. Take the notched trowel and using light pressure, evenly spread the adhesive across the floor. If you can't see your working lines through the ribbons of adhesive, don't cover the lines.

Until you have a feel for how much area you can cover with adhesive and, working comfortably, get tiles installed during its open time, be conservative.

Placing the tiles. The two effective patterns for laying tile are illustrated below. Sequence A is better when you're working with an adhesive with sufficient open time or with self-sticking tiles. Sequence B is better when you have to work quickly.

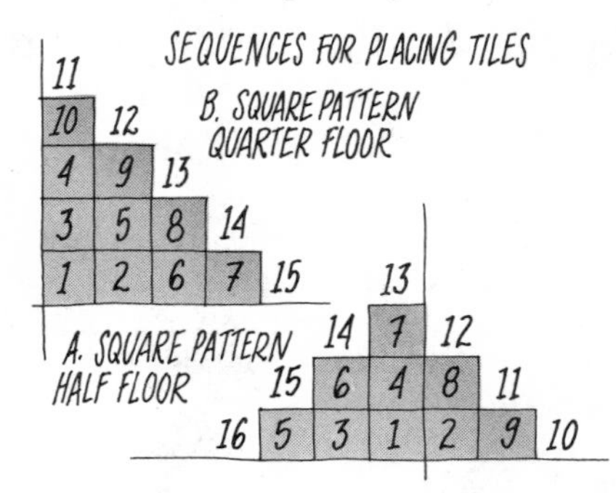

With either pattern, place the first tile at the point where the working lines cross, lining it up carefully so it fits squarely against the lines. Then simply follow the sequence you've chosen.

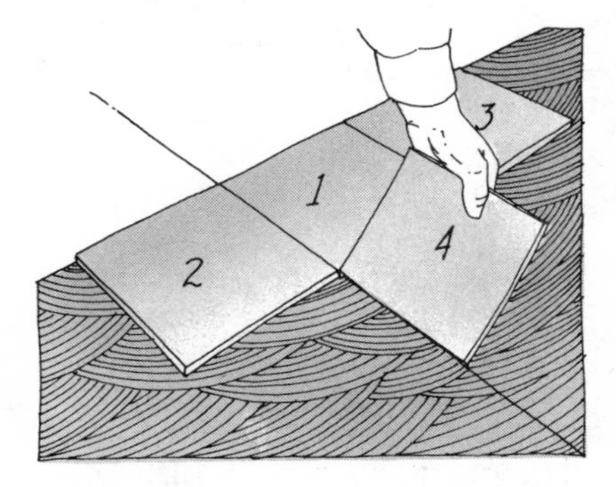

When placing each tile, align an edge with a working line or an adjacent tile, then let the tile fall into place. Never slide the tile into position, or you'll force adhesive up between the tiles. When a tile has been positioned correctly, press it down firmly into the adhesive.

If you're installing solid vinyl or cork tiles, you'll have to bed the tiles in the adhesive with an ordinary rolling pin. But don't roll vinyl-asbestos or asphalt tiles. To avoid nudging tiles out of alignment as you work your way across the floor, use a piece of smooth plywood or hardboard to kneel on.

Many professional flooring installers prefer to place all full tiles first, then go back and cut border tiles and tiles that must be fitted around doorways and other obstacles. Either follow this method, or trim and lay the tiles as you go. Leave about a ⅛-inch gap between the edge of the flooring and the wall.

As you move from one section to another, keep some solvent (recommended by the adhesive manufacturer) handy and wipe up any smears of adhesive as they occur. Adhesive squeezed up between the tiles should also be taken care of as you go, but be careful not to apply too much solvent. Solvent that works its way into the joints can loosen the tiles.

Cutting border tiles. Cut vinyl-asbestos and asphalt tile by scoring the tile with a linoleum or utility knife, then snapping the tile at the point of the cut. Irregular cuts can be made with heavy-duty scissors or shears if the tile is warmed. A hairdryer is a handy tool to use to heat tiles.

The more pliable solid vinyl tiles must be cut with heavy-duty scissors—they don't snap. Vinyl also cuts more easily if it's warmed. Tiles should be warm, but not hot to the touch.

Most cuts necessary to fit tiles around the edges of the room will be simple, straight cuts. A quick method for marking tiles for straight cuts is to place a loose tile squarely on top of the last full tile nearest the wall (see step 1 in illustration at right); then place a second tile on top of it, with one edge against the wall. Draw a pencil line on the lower tile along the edge of the upper tile. Cut the lower tile along the pencil line, and you'll have a piece of tile that should fit the border space exactly. This same technique can be used for cutting L-shaped tiles for outside corners (see steps 1, 2, and 3 in illustration at right).

For irregular cuts needed to fit tile around doorjambs and woodwork, two methods are commonly used. One is to cut a pattern out of heavy paper or

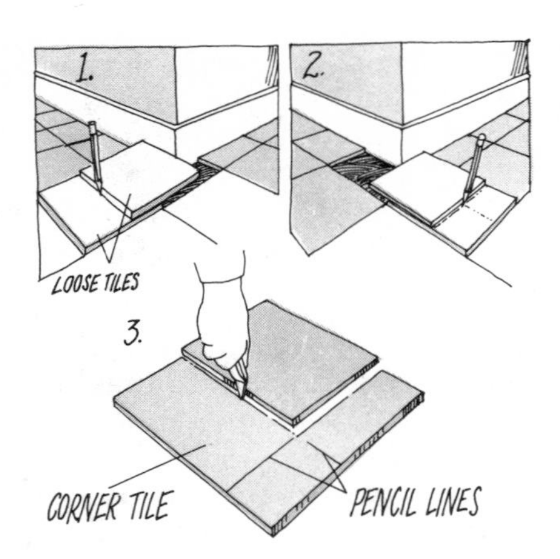

cardboard, then trace the pattern onto a tile. The second is to use a pair of dividers to transfer the irregular pattern directly onto a piece of tile; be sure to align the tile with its "row" first.

Finishing the job. Once all tiles have been laid and border tiles fitted, make a final inspection. Check to see that all tiles are firmly set and smooth. Use solvent to clean up any adhesive that might have been tracked onto the surface or forced up between tiles.

When the floor is clean, replace any baseboards that have been removed and reattach the molding. When attaching the molding, take care to drive the nails into the baseboards, not into the flooring.

Instead of baseboards and shoe molding, vinyl wall base may be used. See page 62 for installation instructions.

CARING FOR RESILIENT FLOORING

The two greatest problems of maintaining resilient flooring are too infrequent cleaning and overpolishing. Grit particles have an extremely abrasive effect on the floor's surface; these particles make thousands of small cuts that dull the surface of the floor. You can prevent surface wear and damage by taking the time to sweep or vacuum the floor regularly—daily, if necessary.

Together with sweeping up dirt, an occasional cleaning with a mild cleaning solution, such as clear or sudsy ammonia, is sufficient to keep resilient flooring in good condition. Use a clean sponge mop for washing and rinsing. Avoid flooding the surface and don't use soap, detergents, or harsh chemicals.

No wax surfaces. Most resilient floor coverings sold today have no wax surfaces. Many remain shiny for years without any polishing; others may require a coat of polish just to add a bit of luster and fill in tiny scratches. Purchase only those floor care products recommended for use on the type of flooring you've installed.

Some products on the market claim both to clean and shine a floor in a single operation. But the labor saved doesn't offset the complications caused by continued use of one-step clean and shine products. In these products, a detergent and a polish are combined; the theory is that the detergent loosens dirt particles, which are retained in the applicator, as the polish is spread.

The system works reasonably well on lightly trafficked floors. In more soiled areas, though, the applicator can't absorb all the dirt, so it's spread and trapped in the polish. In time, this dirt buildup adds damaging abrasive materials to the floor's surface.

Waxing resilient flooring. The surface of vinyl-asbestos tiles and other conventional resilient flooring is porous. You'll want to wax such a surface occasionally to make it shinier and to seal it.

Modern floor waxing products actually often contain no wax. Instead, vinyl or some other plastic is used in the solution. Water-base emulsions are considered safe for all type of synthetic resilient flooring materials currently being produced. Solvent-base wax, which includes paste wax, works well on wood and linoleum, but should *not* be used on vinyl, asbestos, or other resilient flooring materials.

Waxing too often causes the polish to build up and creates a dull, yellow surface. Apply polish only when needed and only to those areas that receive the most wear.

CERAMIC— PLANNING IS CRUCIAL

Installing a new floor of ceramic tile can be a very satisfying do-it-yourself project. Manufacturers offer a wide range of styles and types of tile. New, easy-to-install tiles and improved adhesives and grouts make it possible for a careful, patient do-it-yourselfer to create a tile floor of professional quality.

If ceramic tile is your choice for a new floor, your first task will be to select the tile that suits your needs and taste, the appropriate adhesive for attaching it to the subfloor, and a grout that's right for the type of tile you've chosen and for the room you plan to put it in.

A MYRIAD OF CHOICES

Selecting the type and pattern of tile you like can be difficult, simply because there are so many choices. Visit tile specialty stores or a building supply center with a good flooring materials department.

As you shop around, ask to see only those kinds of tile manufactured for floor surfaces.

You'll find floor tiles available in squares, rectangles, hexagons, octagons, and exotic shapes like ogee and Moorish.

How tiles are finished

No matter what their size, shape, or color, all floor tiles come in one of two finishes—glazed or unglazed. *Glazed tiles* have a hard surface that has been applied between the first ("bisque") firing and the final one. The color is in the glaze. Different glazes can give tiles a high-gloss shine, a satinlike-matte or semimatte look, or a dull, pebbly-textured finish.

Unglazed tiles are naturally dull. Their color comes from the clay they're made of or from pigment mixed with the clay. The color is constant throughout. Samples of unglazed tile that appear to have a glossy finish have been treated with a sealer; some are then either waxed or covered with a glossy floor finish.

The basic types of tile

The four basic kinds of ceramic floor tile are quarry, pavers, patio, and glazed. A fifth type—popular in certain regions of the United States—is Mexican tile.

Quarry tiles are available either unglazed in natural clay colors or with glazed surfaces. These tiles range from 3/8 to 7/8 inch thick, and are made in various shapes and sizes. They have a slightly rough surface and are water-resistant.

Pavers, like quarry tile, come in natural or earthtone colors. They're available glazed or unglazed in several sizes—4 by 4, 6 by 6, and 4 by 8 inches; their thickness varies from 3/8 to 1/2 inch. Pavers are rugged and water-resistant.

Patio tiles, generally thicker and less regular in shape than quarry tiles or pavers, come in reds, tans, and browns. They may be up to 1 inch in thickness and 12 by 12 inches in size. As they can shatter in freezing temperatures, they're best used indoors or, if outside, in warm climates.

Glazed tiles, which include mosaics, are available in many sizes and shapes. Small glazed tiles 1 by 1 inch in size are usually available in sheets held together with a plastic or cotton mesh backing. They're made in literally hundreds of colors and patterns. Glazed tiles often have a glossy finish, but some can be ordered in matte or textured finishes that make them less slippery.

Mexican tile is a terra cotta-like type that has become popular in the American Southwest and West. Some suppliers classify it as a quarry tile and others as a paver. Relatively soft and unglazed, Mexican tile can be protected with a sealer to keep the surface from powdering.

SELECTING AND ORDERING TILE

When buying tile, keep these tips in mind to save yourself time and money:

To avoid long delays, select tile that your supplier has in stock or that can be ordered and delivered within a reasonable period of time.

To make sure you'll have enough tile to complete the job, it's important that you can tell your tile dealer the size of the area you wish to cover. Draw a scale map of the floor area on a piece of graph paper, noting all measurements, so your dealer will be able to determine how much tile you'll need and how much it will cost.

Be generous when estimating and ordering tile. You'll want not only enough tile to complete the job, but also enough extra tile to allow for miscalculations and breakage in cutting trim and border pieces to size. You'll also want to put some tile aside for future use in case a chipped or damaged tile needs replacing.

When you take delivery on your new tile, check each carton to make sure the colors match.

ADHESIVES

Though many professional tile setters insist that tile should be set in the traditional mortar bed, today most tile manufacturers and dealers recommend an adhesive to use with a specific type of tile, depending on your subfloor.

Called "thin-sets" or "thin-beds" (to differentiate them from the thicker cement mortar beds), adhesives have either an organic base, a cement base, or an epoxy base. Each type is applied with a notched trowel.

The characteristics of the three kinds of adhesive, as well as the advantages of a traditional cement mortar base, are discussed below.

Mastics: organic adhesives

Commonly called mastics, organic adhesives come in paste form and are available in two types. Both are suitable for use on concrete or plywood subfloors, or over masonry floors.

Type I mastics are formulated with a solvent. Because they're water-resistant, these mastics are recommended for laying tile in bathrooms, basements, and other damp areas. Since they're flammable and can irritate your lungs and skin, handle them with care and use them only in well-ventilated areas away from open flames. Wear gloves when you work with these mastics.

Type II mastics have a latex base that makes them less irritating to the skin and much easier to clean up than Type I. They're best used to lay tile in rooms relatively free of moisture.

Mortars: cement-base adhesives

These cement-base adhesives aren't flammable and can be cleaned up with water. Mortars are preferred for installing tile over clean, dry concrete slabs and other masonry-type surfaces. Some new mortars are suitable for use over other subfloors, but check with your dealer to be sure.

The two types of mortars are dry-set and latex-Portland cement. Generally, each type is premixed with sand before packaging. If not, follow the directions on the package.

Both dry-set and latex-Portland cement mortars need to stand for 15 minutes after mixing. The mortar should be remixed just before it's applied to the subfloor. The consistency should be just thick enough to form ridges as the mortar is spread with a notched trowel; the mortar should not slump or flow.

Dry-set mortars, in contrast to traditional cement mortar beds, don't require that tiles be soaked in water before being set. Dry-set mortar is mixed with water before it's used; directions are on the package.

Latex-Portland cement mortars are similar to dry-set mortars, except they're mixed with liquid latex before using. This makes the mortar easier to use and more water-resistant than the dry-set type. Directions on the package will give the proportion of latex required.

Epoxy-base adhesives

Though epoxy-base adhesives provide the strongest bond between the tiles and the subfloor, they're also the most expensive type of adhesive. They work well over concrete slabs, masonry floors, old resilient flooring, and wood subfloors—especially plywood subfloors in damp areas.

Some epoxies called "mortars" are useful where their higher resistance to chemicals and their greater bonding strength are important. Ordinary epoxy adhesives have less chemical resistance than epoxy mortars, but have more resistance than other adhesives and are easier to apply than epoxy mortars.

Epoxies are mixed just before application and should be applied when the temperature of the room is between 60°F/16°C and 90°F/32°C. How fast they'll harden, though, depends on the exact temperature, so working with epoxies can be tricky. Smears of epoxy are also difficult to clean off tiles after the adhesive has set. And because epoxies can cause skin irritation, rubber gloves are a must when working with them.

Cement mortar—the traditional adhesive

A mixture of cement, sand, and water, cement mortar is the base used by professional tile setters. Because it's spread in a ¾ to 1¼-inch bed, cement mortar is an effective base to use over uneven concrete slabs or old masonry surfaces. It also works well as a bed for handmade Mexican tile and other irregular or uneven tile.

Consult your tile supplier for instructions on laying a mortar bed suitable for the tile you're installing.

THREE TYPES OF GROUT

Grout is the material used to fill the joints between tiles after the tile has been set in adhesive. The three basic categories of grout are cement-base grout, epoxy-base grout, and silicone-rubber grout. The type you should use will depend on the tile you're installing, where it's located, the adhesive it's set in, and the width of the joints you're leaving between tiles. Ask your tile supplier to recommend the specific type of grout for your project.

Cement-base grouts

Used by both amateur and professional tile setters for finishing most ceramic tile floors, these grouts are made with a base of Portland cement modified to provide water retention, hardness, flexibility, and uniformity. Before buying this type of grout, test it for freshness. If the grains tend to ball together, the grout is stale.

Described below are the four types of cement-base grouts.

Dry-set grouts are similar to dry-set mortars. Generally available premixed, these grouts require you to add water before using.

Latex-Portland cement grout, a mixture of cement and sand, is similar to latex-Portland cement mortar. You need to add liquid latex to this grout.

Sand-Portland cement grout is mixed from fine sand, cement, and water on the job. The proper ratio of cement to sand will depend on the width of the spacing between tiles (see chart below); the more sand, the stronger the grout will be. Up to ⅕ part lime is usually added to make the grout more workable.

Joint width	Cement	Sand
Less than ⅛"	1 part	1 part
⅛" to ½"	1 part	2 parts
More than ½"	1 part	3 parts

Commercial Portland cement grout, sold premixed, is best used with mosaic, quarry, and paver tiles. You must moisten the tile first and damp-cure the grout (see "Applying grout," page 71).

Epoxy-base grouts

The qualities that make epoxy-base grouts effective—their strong bonding properties and water resistance—also make them more difficult to work with than cement-base grouts mixed with water or liquid latex. Epoxy-base grouts must be mixed carefully at the site and are extremely difficult to clean up once they've begun to harden.

Though epoxy grouts are the most expensive, the higher cost and extra work involved in using them may be worthwhile for a heavy-traffic area. Epoxy grouts are available in black, dark brown, gray, and white (the white may turn yellow with age).

Those epoxy grouts specially formulated for use with unglazed tile usually contain a coarse silica fiber that increases bonding strength.

Silicone-rubber grout

Because it stays permanently flexible, repels water, and resists mildew, silicone-rubber grout is often used to seal the seams around bathtubs and other bathroom fixtures. Though too soft for use between floor tiles, it does come in handy for filling the joints between a rigid tile floor and an adjoining floor of wood or resilient material.

TOOLS AND SUPPLIES

The specialized tools required to lay ceramic tile are generally available for rent or loan from tile dealers. The most important tool is a tile cutter, designed to make straight cuts across the face of glazed tiles. Be sure the tile cutter you plan to use has a sharp cutting wheel.

Tiles can also be cut with a glass cutter and a straightedge.

For cutting irregular shapes, a pair of tile nippers is best; or you can use slip-joint pliers instead.

The only other special tools you'll need are a notched trowel for applying adhesive (your tile dealer can recommend the correct notch size), a rubber-faced float for applying grout (a squeegee can be substituted for the rubber-faced float, but it may not be as effective), and a jointer or striking tool for smoothing the grouted joints.

You probably already have the other necessary tools: a putty knife, chalk line, level, folding rule or steel tape measure, a claw hammer, a mallet, and a square.

For the battens that will help you lay straight courses of tile, have on hand some long, *straight* pieces of 1 by 3 or 1 by 4. And to ensure proper spacing between tiles that are uniform in size, you'll need a tile stick (see drawing on page 69) and molded plastic spacers—available from

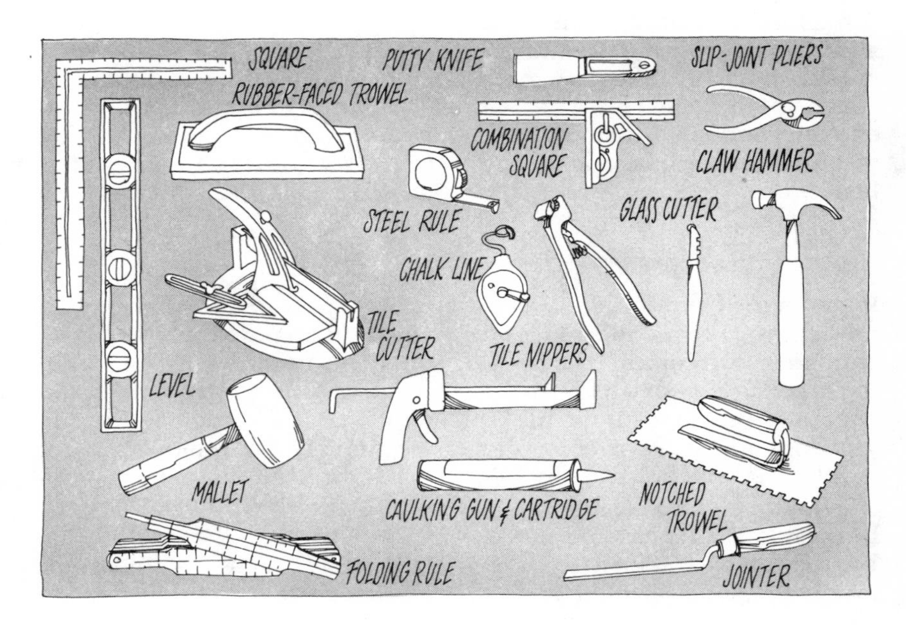

your tile supplier—or uniform spacers cut from wood.

Prepare a piece of wood to use for "beating in" newly laid tiles. Make the piece large enough to cover several tiles and wrap one side with a scrap of thin carpeting or other cushioning material.

For finishing grout and cleaning up, you'll want a sponge and some clean, soft rags. Since many grouts can irritate your skin, have gloves handy.

PREPARING THE SUBFLOOR

Unless it's installed over a perfectly sound subfloor, the most carefully laid ceramic tile floor will eventually begin to deteriorate.

On this and the next page is information on preparing different types of subfloors for a new ceramic tile floor. Before you begin, inspect the subfloor from above and below for any necessary repairs. And don't overlook the supporting structure; loose boards or a seemingly random low spot in the subfloor may indicate a defect in the underlying structure.

Information on examining a subfloor and making repairs begins on page 87. Take time to go through the entire series of checks suggested; it may save you time, frustration, and money later on.

Preparing a concrete slab

If you're drawing up plans for a new home or room addition and want to install ceramic tile flooring, a concrete slab will make the best possible base for the tile. If you already have a concrete subfloor, so much the better.

New or old, the concrete must be completely dry before you can begin preparing it for tile installation. On page 46 you'll find directions for testing a concrete slab for moisture and suggestions for drying.

Once the slab is competely dry, make sure it's clean and free from grease, oil, and old paint or other finishes. Remove grease and oil stains with a chemical garage floor cleaner available at most auto supply stores. Old paint or glossy sealers are best removed with a floor sander equipped with #4 or #5 open-cut sandpaper. (For information on renting and operating floor sanders, see pages 97 and 99.) Vacuum or sweep the floor as you proceed.

If you plan to use cement-base adhesives, fill any holes, low areas, or cracks in the slab with any good concrete patching material. If you're using mastic, you can do the repair work with a mastic underlayment compound.

Finally, scour the slab with a stiff bristle or wire brush. Sweep or vacuum up all loose particles.

Whatever type of adhesive you plan to use, check the directions on the container when you're preparing the slab. Some manufacturers may recommend using a sealer before applying the adhesive.

Preparing a wood subfloor

You can lay ceramic tile over a subfloor made of plywood panels or individual boards if the subfloor is rigid and structurally sound.

Preparing a plywood subfloor. To check a plywood subfloor, make sure that all panels are securely attached to the joists, and that there are no protruding nails. Using a wide-bladed putty knife or similar tool, remove plaster spills or other imperfections. Minor indentations, holes, or cracks will be filled with adhesive as you install the tiles.

If the plywood panels are so lightweight that they move when you walk on them, reinforce the floor with a second layer of plywood. Use 4 by 8 or 4 by 4-foot panels of exterior or underlayment-grade plywood; the panels should be at least ⅜ inch thick if you're using mastic, ⅝ inch thick for epoxy adhesive. Stagger the second layer of panels so the joints don't fall directly over the joints in the bottom layer. Avoid laying panels so four corners meet (see illustration, page 48). Leave about ⅛-inch gaps between panels.

Fasten down the new plywood with 6-penny ring-shank nails spaced 6 inches apart. Begin nailing in the center of each panel and work toward the edges. Where possible, drive nails through the panels into the joists.

Preparing a wood board subfloor. If the subfloor is made of individual 4 or 6-inch boards, make sure that each board is securely attached and that any protruding nails are driven flush. To prevent the floor from warping, you'll have to cover a board subfloor with a layer of plywood before installing ceramic tile (see above).

Strip, plank, or wood block floors in good condition can also be covered with ceramic tile. For instructions on making minor surface repairs in wood floors, see "Floor Surfaces—Repairs that Work Wonders," page 101. If the old floor is sound and level, you may only need to give it a rough sanding with a floor sander to remove the old finish and smooth out any rough areas.

Before laying ceramic tile over a wood floor, be sure to check the adhesive you're using for any special instructions.

Preparing an existing floor

Whenever possible, old flooring should be removed before installing new ceramic tile flooring. Not only is it easier to examine the subfloor and make any necessary repairs, but also the new floor will be level with the floors in adjacent rooms. If these two factors aren't a concern for you, you can lay the new tile directly over the old floor.

Well-bonded resilient flooring, if it's level and in good repair, can be successfully covered with tile. To make minor repairs on resilient flooring, see "Repairing minor surface damage," page 107. Cushioned resilient flooring (sheet or tile) is too springy to be used as a base for ceramic tile and must be removed. For tips on removing old resilient flooring, see "Old resilient flooring," page 49.

Normally, an organic or epoxy-base adhesive is used to install tile over old resilient flooring. Your tile supplier can recommend the best adhesive to use, depending on the type of old resilient flooring you'll be covering.

And don't overlook the possibility of covering an old floor made of ceramic tile, flagstone, slate, or other masonry flooring material. The old surface must be level, clean, and dry, and any loose pieces must be secured. Your tile dealer can recommend the adhesive you should use.

Whatever kind of old floor you plan to cover, you'll have to take up the shoe molding before putting in the new floor. Baseboards should be removed only if they can be pried loose without damaging walls and door casings. As you remove molding or baseboards, number the pieces with chalk or a pencil so you can replace them in their original positions after the new floor is in place.

LAYING THE TILES

On this and the following pages, you'll find step-by-step instructions for laying ceramic tile in thin-set adhesive. If you're using a cement mortar bed, ask your tile dealer for instructions on laying the bed and placing the tile.

Checking tools and supplies

Before you begin, make sure you have everything you'll need for the job. Open and inspect the cartons of tile—is the tile the right type and color, and is there enough to cover the floor? Read the directions on the container of adhesive to be sure you have the proper adhesive and solvent in the quantities you'll need.

If the tile is dusty, wash and dry it before installation. Even a little dust can prevent adhesives from forming a strong bond.

Establish working lines

The key to laying a floor with straight rows of tile parallel to the walls is to lay out proper working lines. You can begin either at the center of the room or at one wall, unless you're laying tile in a cement mortar bed. Then you'll have to start at one wall to avoid disturbing the carefully leveled bed as you work.

Working from the center of the room. In this method, you begin by laying out working lines that intersect at a right angle at the approximate center of the room. Starting in the center of the room will make it easier to keep the rows even and is the best method to use if the room is out of square or if you've chosen a tile with a definite pattern or design.

For instructions on laying down working lines in the center of a room, see "How to Establish and Use Working Lines," page 54. After you've established the lines, nail batten boards along them to provide rigid guides for the first course of tiles—but only if you're working over a wood or resilient subfloor. Over a concrete slab, you'll have to use the chalk lines as your guide.

You start setting tile where the working lines cross, filling up the floor quarter by quarter. Work on the quarter with the doorway last.

Starting at the wall. Use this method only if two adjoining straight walls meet at an exact 90° angle. To check for square corners and straight walls, place a tile tightly into each corner. Stretch a chalk line between the corners of each pair of tiles, pull the line tight, and snap each line.

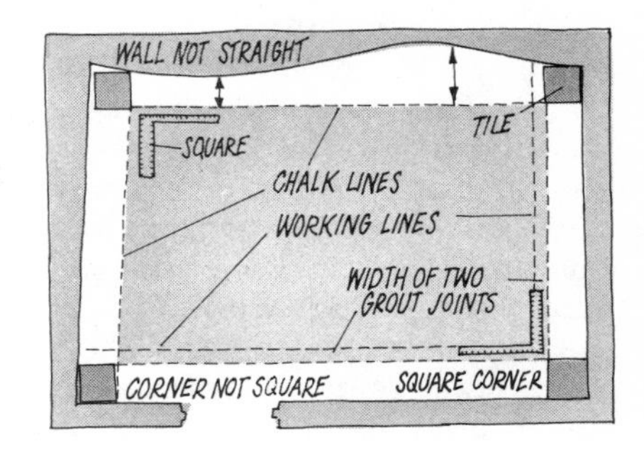

Variations in the distance between the chalk lines and the walls will show any irregularities in the walls. You can ignore small variations—about the width of a grout joint. With a carpenter's square, check for square where the lines intersect in each corner of the room.

It's a good idea to make a dry run before you begin setting the tiles in adhesive. Lay the tiles out on the floor, allowing the proper spacing for grout joints; using a tile stick as shown in the illustration will help you achieve uniform spacing. This way, you can determine the best layout of tiles and keep the number of tiles to be cut to a minimum. You may

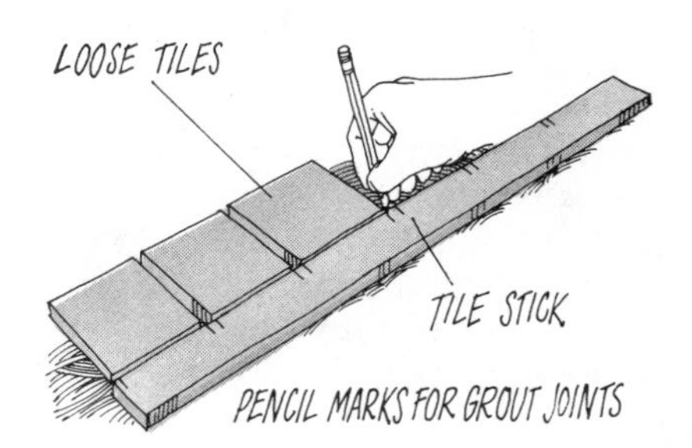

find that a slight reduction or enlargement of grout joint widths will make it possible to set a row of tiles without any cutting.

You can start at any straight wall adjoining a square corner. Snap a new chalk line parallel to the first line and approximately two grout joint widths away from it toward the center of the room. Lay a similar line, at a right angle to the first, along an adjoining wall (see illustration on previous page).

Next, nail a batten (see illustration below) or wood straightedge along each of the new working lines. The two battens should form a right angle; if they don't, check your measurements and adjust the working lines accordingly. If you're working over a concrete slab, toenail short wood spacers at right angles to the batten boards to brace them securely against the walls.

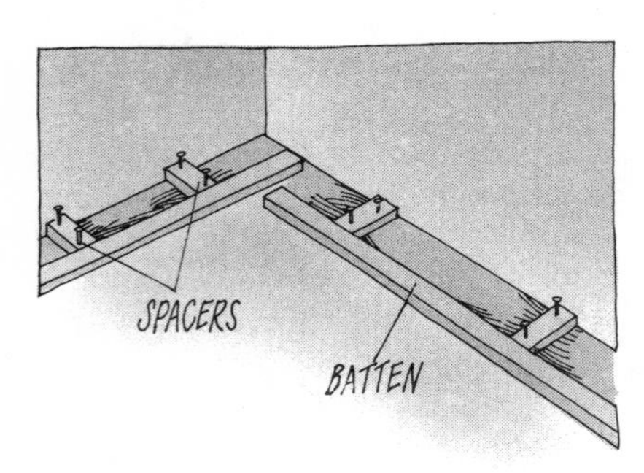

Setting the tiles

Read through the instructions that follow before you begin setting tiles in adhesive, and be sure you understand all the steps involved.

To begin laying tiles, start spreading a strip of adhesive on the floor along one of the battens, using the notched trowel; cover about a square yard at first, the area you can comfortably tile before the adhesive begins to set. As you become more proficient, you may be able to spread the adhesive over a larger area.

Using a gentle twisting motion, place the first tile in the corner formed by the two battens. With the same motion, place a second tile alongside the first. To establish the proper width for the grout joint, use molded plastic spacers (available from your tile supplier) or uniform spacers cut from wood. Remove the spacers when the adhesive begins to set.

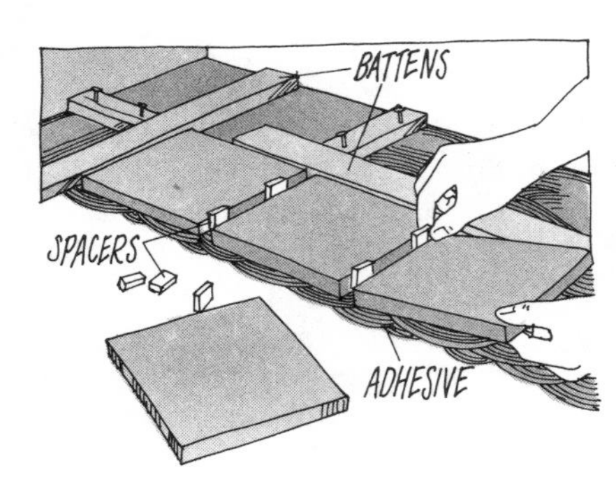

As you work, clean off any adhesive from the surface of the tiles, especially if you're using an epoxy; after it has set, it cannot be removed from the tile surface.

If you're working from the center of the room, lay tiles and spacers following one of the patterns shown on page 54. If you've started from one wall, set tiles in continuous rows, beginning each row at the same wall.

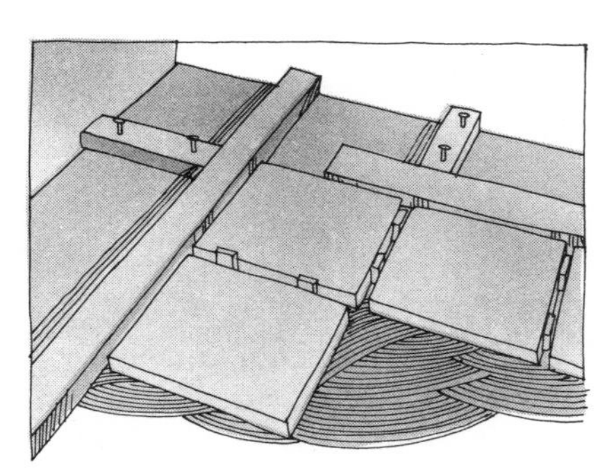

When you're ready to install the border tiles. remove the battens. Then mark the border tiles for the necessary cuts. To mark simple, straight cuts on tiles, see page 64.

Tiles can be cut with a tile cutter rented from your tile dealer or with a glass cutter. *Be sure to wear safety goggles when cutting tile.*

If you're using a tile cutter, score the finished surface along the pencil line with the cutter; then press down on the handle to break the tile (see illustration). Or you can score the surface with a glass cutter and straightedge (see drawing). To break the tile, place the score line over a dowel or nail and press down evenly on both edges.

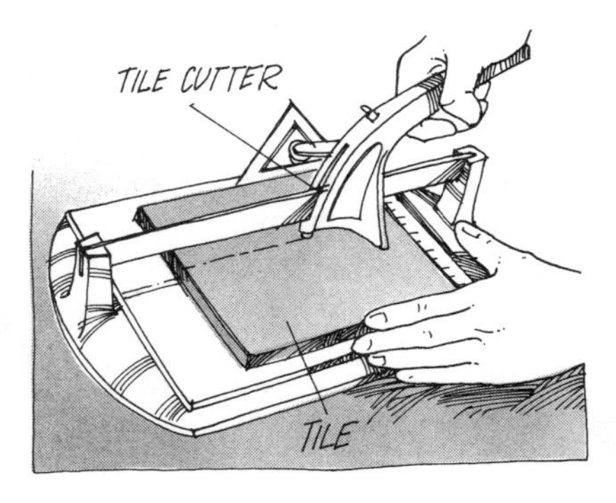

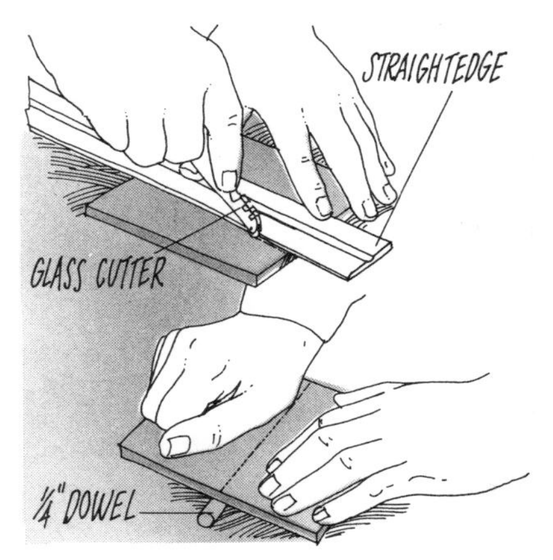

To cut irregular shapes, use a tile nipper. Scoring the cutting line with a glass cutter first helps.

Smooth any rough or jagged edges by rubbing them against a whetstone with water.

As the tiles are laid, set the piece of carpet-wrapped wood (see page 68, top left) over the tiles and tap it with a mallet or hammer. This process, called "beating in," beds the tiles in the adhesive and ensures that they'll be level.

If possible, avoid walking on newly laid tiles until the adhesive has thoroughly set. If you must work over them to complete a quarter-floor pattern, put down plywood to spread your weight over several tiles.

From time to time, use a carpenter's square and a straightedge to make sure each course is straight. If any tiles are out of line, wiggle them back into position while the adhesive is still flexible. Minor irregularities can become major misalignments if they're not corrected promptly.

When all the tile has been laid, make sure you've removed all the spacers; then clean the tile surface so it's completely free of adhesive. Check the joints between tiles; remove any adhesive that would make the joints too shallow for holding grout.

Allow the tiles to set properly before applying grout. Setting takes about 48 hours for tiles laid in cement mortar, 24 hours with mastics, and 16 hours with epoxy adhesives. Since the ungrouted tiles can break very easily, keep all unnecessary traffic off the floor.

Applying grout

Most grouts are applied in the same way. Grout is worked into the open joints; the excess is removed; then the tile is cleaned and the grout given time to cure.

Use silicone-rubber grout in joints where the tile floor meets an adjoining floor of a flexible material such as wood. Unlike other grouts, silicone rubber is applied with a caulking gun.

Grout can be applied liberally around glazed tiles; but grouting unglazed tile requires more care, as the grout can stain the unsealed surface of the tile. Whether your tile is glazed or unglazed, be sure to read the directions and recommendations of the grout manufacturer before you apply any grout.

Use a rubber-faced float or a squeegee to apply grout. Have on hand water or the specific solvent recommended by the grout manufacturer, and some soft rags for cleaning up as you work. Wear gloves to protect against skin irritation.

Spreading with a float, apply grout to the surface of the tile.

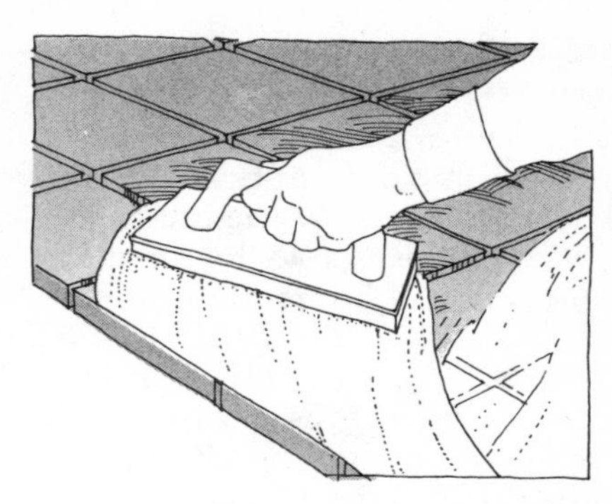

Force it into the joints so they're completely filled; check that there are no air pockets. Scrape off the excess grout with the float, working diagonally across the tiles.

Soak a sponge in clear water and wring it out. Using a circular motion, wipe the tiles, removing any remaining grout, until the joints are smooth and level with the tiles. Rinse and wring out the sponge frequently. When the tiles are clean, let the grout dry for about 30 minutes.

Any film of grout left on the tile surfaces will form a soft haze when dry. Polish it off with a soft cloth. You can leave the grout with the rough surface created by the sponge or use a jointer or

CARING FOR A CERAMIC TILE FLOOR

Ceramic tile flooring can be classified as either glazed or unglazed. Glazed tile is sealed as part of the manufacturing process. Naturally dull unglazed tile has an open, porous surface that's sealed after it's installed.

A number of products designed to produce a hard surface are available to seal unglazed tile. When choosing a sealer for your new tile floor, consider both the type of tile you've installed and the look you want to achieve. Some sealers, used together with a floor finish or wax, will finish the floor to a high gloss; others will leave it in its natural form, without any luster at all.

For day-to-day care of a sealed tile floor, just sweep and damp-mop with a mild detergent or all-purpose household cleaner. It's important to sweep away any grit that can be ground into the floor and damage the tile.

If grout becomes stained, scrub it with a toothbrush and household bleach or a commercial tile cleaner. Colored grout may react to bleach, so test a small area before attacking the entire floor.

Under ordinary use, ceramic tile doesn't need to be cleaned often. A light coat of wax (a type specifically recommended for tile floors) will bring out the color of the tile and add extra protection. Follow the manufacturer's directions for applying the wax.

striking tool to smooth the joints.

Most grouts take at least 2 weeks to cure. You'll want to damp-cure cement-base grout by covering the newly installed floor with plastic; leave the plastic in place for 24 hours; then remove it and allow the grout to cure thoroughly. Put thin pieces of plywood over the floor to avoid stepping on the newly grouted tiles.

Finishing the job. After the grout has fully cured, wash the floor with detergent or household cleaner and water. Let it dry for several days.

Seal the grout with a silicone or lacquer-base sealer as recommended by your tile supplier.

If you removed any shoe molding before installing the new floor, replace it. Drive nails through the molding directly into the baseboards. Don't attempt to angle nails down into the floor.

MASONRY—BULKY BUT WORTH THE EFFORT

Products of the earth, natural masonry materials come in an array of textures, subtle shades of color, and a variety of shapes that can be used to create a striking interior floor. Masonry is made from stone or clay. Whether used in its original form—marble or slate, for example—or in manmade versions, like brick or terrazzo, masonry makes an exceptionally tough, durable floor.

Stone masonry flooring—marble, slate, limestone, granite, and flagstone—is usually installed only in entryways, kitchens, and bathrooms, or used as decorative flooring around fireplaces.

Brick, a relatively inexpensive type of masonry, is becoming an increasingly popular choice for interior floors. Its mass and its heat-retaining property make it ideal for passive solar home designs.

THE TWO TYPES: BRICK AND STONE

With their regional characteristics, masonry products come in a virtually limitless variety. Still, natural masonry can be purchased in one of two basic forms: either as manmade bricks or as stone.

Brick: Manmade masonry

Though most bricks can be used for both walls and floors, bricks intended specifically for floors, whether inside or outside, are called "pavers."

Pavers are available in regular or slightly less-than-regular thicknesses, or as splits—half as thick as regular bricks. Full pavers can be laid on a concrete slab; splits require a concrete slab topped with a mortar bed or spread with thin-set adhesive.

Stone: Roughhewn or finished

Stone selected for interior flooring should be dense, with a hard surface.

Roughhewn stone is always set in a thick mortar base over a concrete slab. The mortar compensates for variations in thickness and provides a solid base for the irregular undersides of the stones.

Marble, slate, or similar natural stone can be ordered in uniform thicknesses cut into squares, rectangles, or irregular shapes.

Relatively thin stone of uniform thickness can be laid in thin-set mortar in a regular, planned pattern. Because this precut material makes for a lighter floor, it requires less structural support than roughhewn stone flooring set in a full bed of mortar.

TOOLS AND SUPPLIES

The specific type of masonry floor you're installing determines the tools and supplies you'll need for the job. Most of the tools are designed especially for masonry work (see illustration).

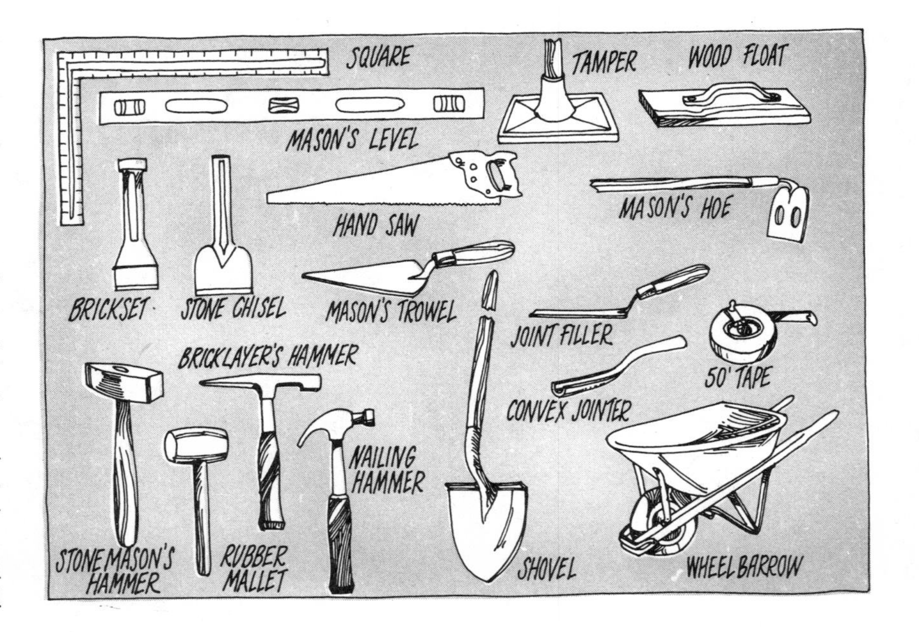

Uniformly cut stone is installed using the same tools and techniques required for installing ceramic tile. To lay brick and irregular stone in mortar, you'll need a trowel, brickset or stone chisel, level, mason's line, stonemason's hammer, convex jointer, and measuring tape. Use a wheelbarrow and mortar hoe to mix small batches of mortar.

PREPARING THE SUBFLOOR

Preparing a base for masonry flooring is relatively simple—if the flooring is to be laid over a concrete slab at ground level. Putting masonry over a suspended floor in a frame home can be complicated because of the weight of masonry.

If masonry flooring was planned as part of the original building, allowances will have been made for the extra weight and the support system beefed up accordingly. But if brick or stone flooring is being laid in an older home as part of a remodeling project, the old floor may need reinforcing.

Always consult a building design professional before putting masonry flooring over any subfloor other than a ground-level concrete slab. Since rigid masonry flooring demands a firm, unyielding base, the subfloor below must be structurally sound (see page 87).

If you're planning to install masonry flooring in a mortar bed over a concrete slab, make sure the slab is free from dust or grease that could affect its bonding with the mortar. The mortar bed will take care of most minor irregularities.

WORKING WITH BRICK

Bricks can be set tightly together with no mortar joints (the most common method used for interior floors, and by far the easiest for the amateur) or set in mortar. Over the years, dozens of brick patterns have evolved; the illustration below shows the most common ones.

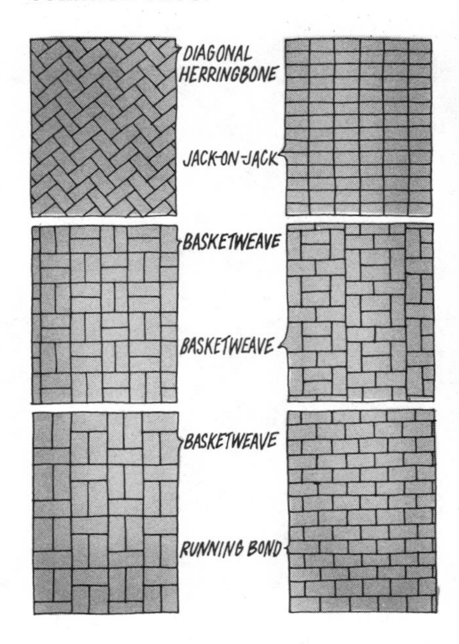

Half-bricks or splits require a rigid foundation for stability. Splits can be used where the extra weight of full-thickness bricks would cause structural problems—on a wood subfloor, for example (see illustration above at right).

Mortarless brick flooring

Ideally suited to the amateur, mortarless brick flooring is quick and easy; it produces a beautiful brick floor with a minimum of fuss.

The drawing at the top of the next column shows how it's done. Mortarless brick flooring can be laid either on a wood subfloor or over a concrete slab. With a new subfloor, the joists should be spaced about 25 percent closer together to handle the extra weight (about 15 pounds per square foot). Joists should have 12-inch instead of 16-inch centers. Existing subfloors can be reinforced, but be sure to consult your building department for exact specifications. Bricks laid over a wood subfloor should not exceed 1⅝ inches in thickness; over concrete slabs, any thickness will do.

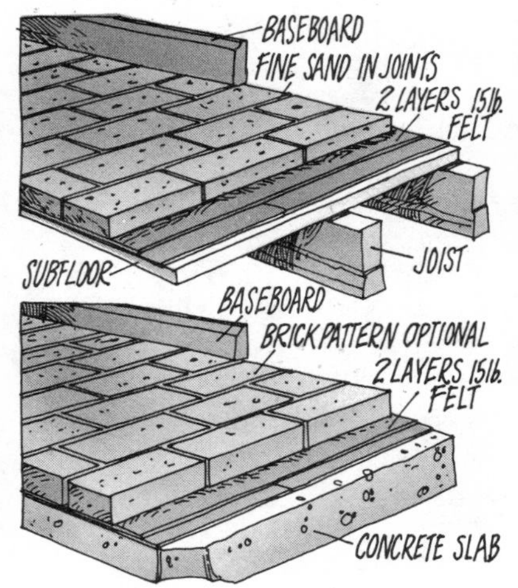

With either a wood or concrete subfloor, begin by placing two layers of 15-pound building paper (felt) over the surface. Butt, don't overlap, the edges, and run the layers at a right angle to each other.

If you're lucky enough to have a truly straight wall, you can use it as a guide for your bricklaying. Otherwise, stretch a line just less than one brick length or width away from the starting wall, depending on the pattern you're using; measure at the ends (for tips on laying out this line, see "How to Establish and Use Working Lines," page 54).

Lay the first course of bricks to this line. After the rest of the bricks are in place, you'll go back and cut bricks to fit between the first course and the wall (see illustration).

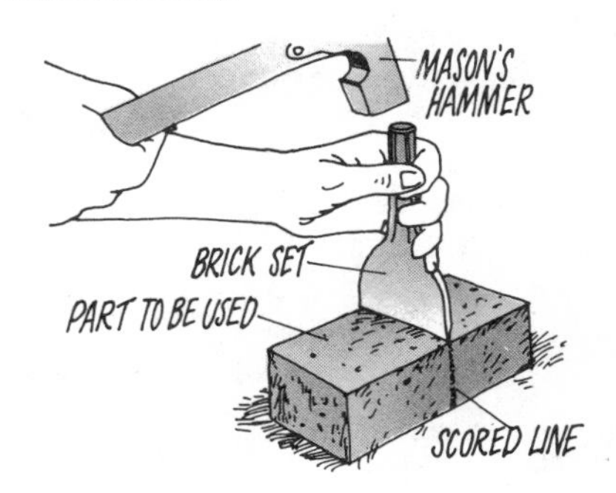

Resetting the line for each course or two as you proceed yields the best results. You can adjust joints between bricks slightly to keep the courses straight; just remember a tightly butted joint is ideal. As you can walk on the floor as it's laid, starting and ending points are not important.

When you've laid all the bricks, spread fine sand over the surface and sweep it diagonally into the joints with a soft brush. It's a good idea to repeat this sanding once or twice at intervals of a few days.

When you're sure the floor is completely dry, you can apply several coats of a masonry sealer. This locks in the sand and gives a slight gloss to the floor.

Mortared brick flooring

A concrete slab provides the best possible subfloor for mortared brick flooring. It's easiest to lay interior brick floors with closed joints (the bricks butted together). Mortared joints call for more expertise but are handsome, and often worth the extra effort.

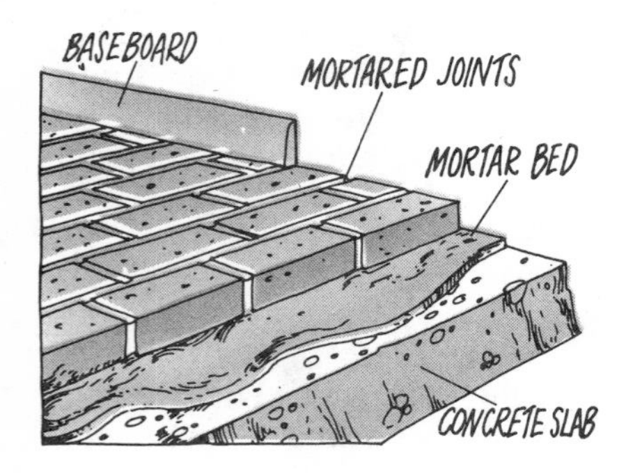

If the slab is not level or has any major imperfections, a mortar bed will compensate. It needn't be thick—½ to 1 inch will do—because it serves no purpose other than to provide a good bond between the slab and the brick.

But keep in mind that laying brick in mortar over a concrete slab is a project that requires a great deal of skill. For amateurs, mortarless brick flooring is a much better choice (see page 73). If your slab is relatively level, you may also want to consider laying a brick floor in thin-set adhesive; just patch any minor imperfections with a commercial compound and proceed according to the directions for ceramic tile beginning on page 69.

Preparing the slab. Before you begin placing bricks, take time to clean the slab thoroughly. Remove any dirt or grease that might prevent a good bond between the slab and the mortar. Wash the slab using a stiff bristle brush and, if necessary, detergent. Dampness should present no problem, as long as the slab has cured fully.

Mixing mortar. It's not difficult to mix mortar in small batches. In any case, you shouldn't mix more than you can use in an hour. A common recipe for mixing a mortar bed is 3 parts graded sand, 1 part cement, and ¼ part lime.

Using either a wheelbarrow or a mortar box, mix the dry ingredients thoroughly; then add the water gradually, mixing it in until the mortar is smooth, uniform, and granular. Mortar should spread easily, like soft butter, without slumping or losing its shape.

Placing the bricks. Plan to work toward a doorway or other exit, and toward your supply of brick and mortar. Stretch a line set about a brick length or width, depending on your pattern, from and parallel to the wall as a guide for laying a straight first course (see "How to Establish and Use Working Lines," page 54). Lay the first course of bricks dry, marking their proper positions in pencil on the slab and allowing for any mortar joints.

Soak the bricks in water for a few minutes. Then, using a wood float and taking care to see that the thickness of the mortar bed remains constant, spread out a layer of mortar over a small section of the slab. Set the first brick in place near the center of the line and work out from that point in each direction, placing the bricks in the pattern you've chosen (see page 73). If you're planning to have mortared joints, use a wood spacer (¼, ⅜, or ½ inch, depending on the brick size) between each brick (see illustration).

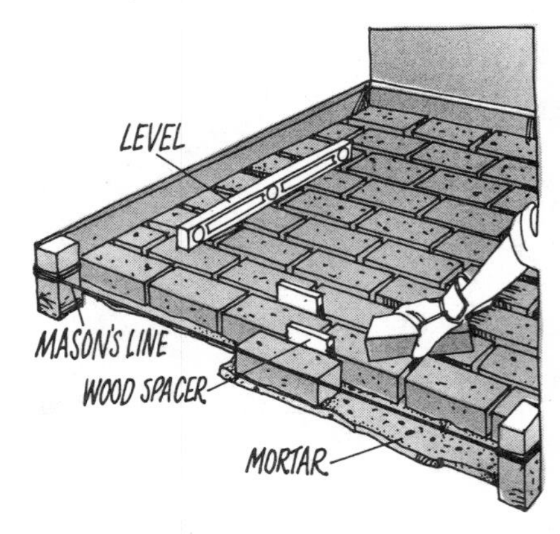

As you set each course, use a level and a line to make sure the surface is level. When you reach walls or other obstructions, cut bricks as needed to trim out the floor—cut pieces must be placed as you go because you won't be able to walk on the newly laid bricks for at least 24 hours.

No matter how carefully you handle bricks and mortar, minor spills are unavoidable. You'll save time and effort by cleaning up as you go. Wipe up any mortar on the surface of the bricks with a piece of wet burlap.

Mortaring the joints. When the mortared brick has set for 24 hours, you can begin mortaring the joints. Mix a small batch of mortar to the same consistency you used to lay the bricks. Working carefully to minimize spills, use a small trowel to pack mortar into the joints. When the mortar is "thumbprint hard," tool each joint with a convex jointer,

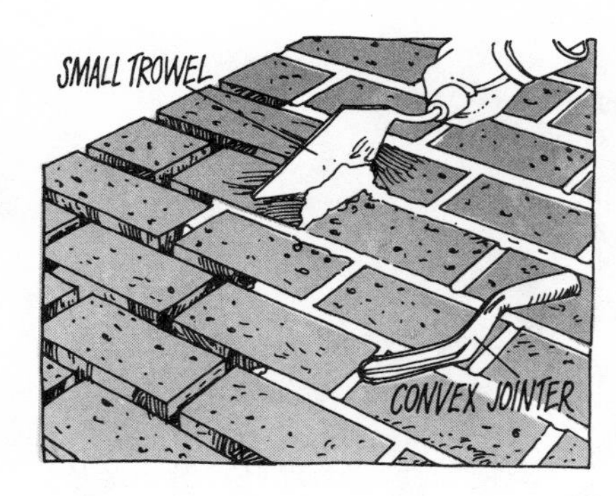

a dowel, or other convex object. Wait several hours; then use a burlap sack to remove mortar tags and stains.

Finishing up. After the final bricks have been placed (and the joints mortared), wait until the mortar has stiffened enough to stand some light traffic. Then make a final check for smears that need to be cleaned up.

If any stains prove stubborn, carefully scrub the area with a solution of trisodium phosphate and laundry detergent (about ½ cup of each to a gallon of water); *do not* let the solution run over the edges of the bricks into the mortar joints.

Sprinkle water over the new floor and cover it with a layer of polyethylene film (4-mil is adequate) to keep the mortar from drying too quickly. Leave the film in place for about 2 days.

Allow the floor to dry completely—this may take several weeks—before sealing the bricks. Any moisture trapped in the bricks when they're sealed can discolor the surface and spoil the seal.

WORKING WITH CUT MASONRY

Granite, limestone, marble, and terrazzo (marble chips suspended in a cement mix and polished to a glasslike finish) are typically used in uniform-size pieces for flooring. These and other stone materials cut uniformly thick can be installed using the same technique, adhesive, and grout (where necessary) used to lay ceramic tile (see pages 65–71).

The initial cost of cut stone is high, and installation mistakes can be expensive, so review the instructions carefully. Also check with your supplier for recommendations on the specific adhesive and grout required for the type of cut masonry you want and the subfloor you'll be covering.

WORKING WITH ROUGHHEWN MASONRY

Slate, flagstone, and other types of roughhewn natural stone are laid in a thick mortar bed for very practical reasons. The irregularity of the material makes it necessary to provide a cushion of mortar that will compensate for differences in thickness from piece to piece. Because roughhewn masonry must be set with open joints, a sand base that might be practical for an outside patio is not workable as a base for an interior floor.

The weight of irregular masonry flooring requires it be laid over a concrete slab. Clean the slab thoroughly to make sure that the mortar will bond properly.

Installing roughhewn masonry units requires a bit of inventiveness, since each piece is a different shape and thickness. You must spread mortar as you go, varying the thickness of the bed to keep the surface level (see instructions below). Units must be trimmed to produce a satisfactory pattern with relatively uniform joints.

With roughhewn masonry, it's essential to have a helper who can keep you supplied with mortar and help you move any large stones.

Planning the pattern

Though it's impossible to lay out a full floor of roughhewn masonry in advance, you can set out large sections dry before committing yourself, and the flooring, to a mortar base.

Distribute the larger pieces in a balanced pattern, and use smaller units or trimmed pieces to fill in the gaps. Once a section is bedded in mortar, move on to a new section, trying a dry fit as you did the first.

Preparing the mortar

Small batches of mortar can be mixed easily by hand in a wheelbarrow or mortar box with a mortar hoe. The basic ingredients—3 parts sand, 1 part cement, and ¼ part lime—should be mixed dry. (Prepackaged mortar mix is a good choice for small projects; its convenience makes up for its greater cost.)

Add water slowly until the mixture is soft enough to work with a trowel, but stiff enough to support the weight of the masonry. The mortar should contain less water than mortar for brick because stone is less absorbent.

Laying the stone

Pick a corner of the room opposite your source of masonry and mortar; you'll work toward your supply and, of course, an exit. Your slab should be slightly damp when you begin; sprinkle it, if necessary.

Using a wood float or trowel, spread mortar evenly over a small section of the floor. Don't cover more area than is required to set two pieces of stone. Tap

each stone level with a rubber mallet or the handle of your trowel, and check frequently with a level to make sure each stone is level with its neighbors. Leave the joints between stones unfilled; you'll grout them later.

As you go, trim the stone to fit along the walls and other obstacles. To trim a stone, lap it over its neighbor (see illustration) and mark the trim line. To cut the stone, score it along the line, top and bottom, with a brickset or stonemason's chisel, prop the edge to be cut on wood, and strike the scored line with the brickset (or stonemason's chisel) and hammer (see illustration). Minor trimming can be done with the stonemason's hammer alone.

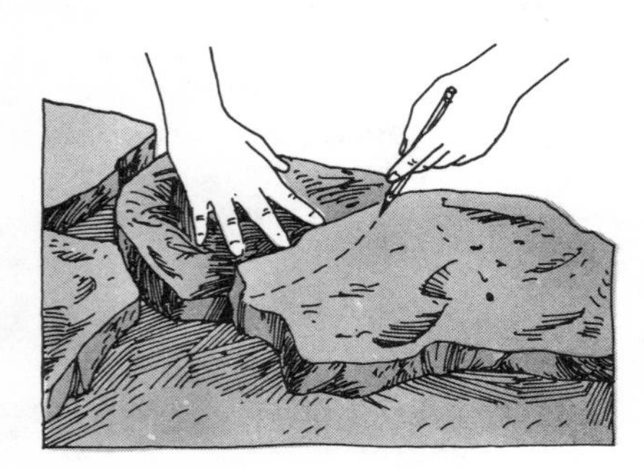

Using a sponge and water, clean up mortar spills as you work. Generally, it's easy to wipe mortar off hard-surfaced stone. Don't plan to use an acid wash later, since acid can stain stone and is impractical to use indoors.

When the stone is laid, let the mortar set for 24 hours. Then prepare a grout of 3 parts sand and 1 part cement (no lime) and using a trowel, fill in the joints between units (see drawing). Or you can make the grout soupier than mortar—about the consistency of a milk shake—and pour it from a bent coffee can into the joints. Wipe up excess grout with a wet sponge.

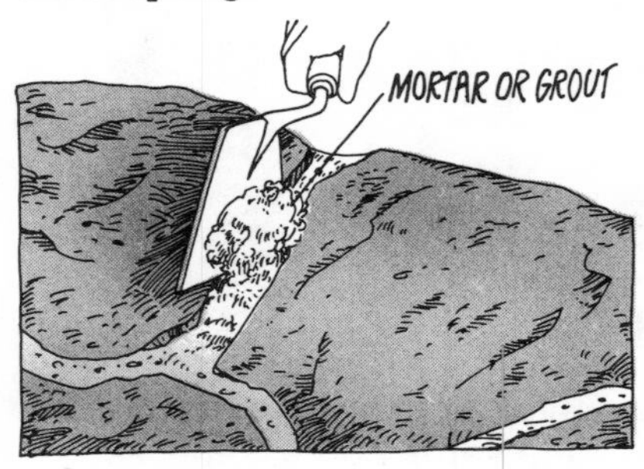

As the grout sets, smooth the joints with a joint filler or trowel, leaving the grout flush with the surface of the stones.

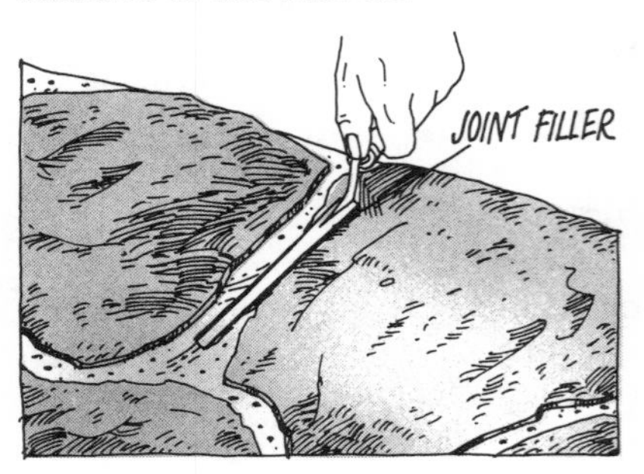

Finishing the job. When the floor has dried thoroughly, apply a dressing or sealer. Ask your building materials supplier to recommend one for the type of stone you've installed.

CARING FOR A MASONRY FLOOR

The porous surface of most masonry flooring must be sealed after the flooring is installed. Special sealers are available for specific types of masonry. Most are penetrating sealers that soak into the porous surface; depending on the flooring material, you may need to apply several coats to produce a hard surface.

Your masonry supplier can recommend the right product for the type of masonry floor you've installed. To apply the sealer, follow the manufacturer's directions.

Once the surface has been sealed, masonry floors are very easy to maintain. The materials are rugged and durable and need only an occasional sweeping and washing with a damp sponge mop and mild detergent.

You may want to apply a light coat of wax now and then to bring out the character and texture of the flooring material. Check the label of the wax container to make sure the wax you've selected is compatible with the masonry material. Generally, you'll be looking for a water-base emulsion wax.

CARPETING—FOR CLASSIC COMFORT

Carpeting woven from natural wool was the standard of quality for many years. No other type of fiber could match wool for warmth, wear, and appearance. But the relatively recent development of manmade fibers and high-speed manufacturing techniques have pushed nylon, acrylic, polyester, and polypropylene fibers into the forefront, and replaced woven carpeting with tufted carpeting.

Though wool is still desirable, synthetics have proven to be long wearing, good-looking, and much less expensive. Moreover, they're easier to handle and less likely to be damaged during installation than wool.

TWO BASIC TYPES: CONVENTIONAL AND CUSHION-BACKED

Regardless of the type of fiber used, synthetic carpeting is manufactured either with or without a rubber backing. Carpeting without a backing is commonly referred to as conventional; carpeting with a bonded rubber backing is called cushion-backed.

Conventional carpeting is usually installed under tension with stretchers and secured around the perimeter of a room with tackless strips. It can also be installed using double-faced tape or simply laid loose, like an area rug, but the carpet edges should be bound to keep them from fraying. Conventional carpeting usually requires a separate pad, not only to keep it from wearing more quickly, but also to cushion the surface.

Cushion-backed carpeting is easier to install than conventional carpeting. No stretchers or tackless strips are required. This carpeting is either laid in latex adhesive or secured to the floor with double-faced tape. Cushion-backed carpet is often less expensive than conventional carpeting, but it won't last as long as good-quality conventional carpeting.

ORDERING CARPETING

Careful planning before ordering carpeting will avoid the complications of ending up either with carpeting that doesn't fit properly or with the extra expense of ordering more than you need.

Drawing a floor plan. Prepare a scale drawing of the area to be carpeted, noting exact measurements and marking doorways and other obstacles.

These measurements will help you and your carpeting supplier select the carpeting that fits your particular floor plan most economically. You'll also need the floor plan to cut the carpet later on.

Locating seams. Selecting carpeting will be easy if you're fortunate enough to be able to cover your floor with a single piece. But if you'll need seams, here are a few special points to consider in locating them.

Because seams are the weak points in wall-to-wall carpeting, you'll want them where they're the least visible and away from the room's most heavily trafficked area. Seams also tend to be less visible when they run in the direction of a room's primary outside light source.

Once carpeting is made into rolls, the tufts, regardless of whether they're sculptured, plush, or looped, will fall in one specific direction. Since carpeting generally will look the richest and show its color to best advantage when you look into the pile, try to have the pile pointing in the direction from which it will be viewed most often—toward the doorway, for example. All the pieces must run in the same direction, also.

And don't overlook the need to match patterns. Special allowance has to be made in figuring the right amount of carpeting in this case.

Estimating padding. All carpeting feels more comfortable underfoot if it has padding. But even more important, padded carpeting wears longer.

Typically, padding presents few problems in ordering. To come up with the most economical buy, ask your carpet dealer to use your floor plan to figure out exactly how much padding you'll need.

TOOLS AND SUPPLIES

The hand tools required to install either conventional or cushion-backed carpeting include a screwdriver, hammer, tape measure, straightedge, chalk line, and heavy-duty shears.

Special tools for installation. Depending on the type of carpeting you've selected and the installation method, you'll need some special tools, most of which can be rented from your carpet dealer.

Conventional carpeting installed under tension requires the use of carpet stretchers. For close work, a knee-kicker is used. A power stretcher, equipped with 20-foot-long extensions, is used to stretch carpeting across a room.

Other tools you'll need to install conventional wall-to-wall carpeting are a wall trimmer, seaming iron, row-running or

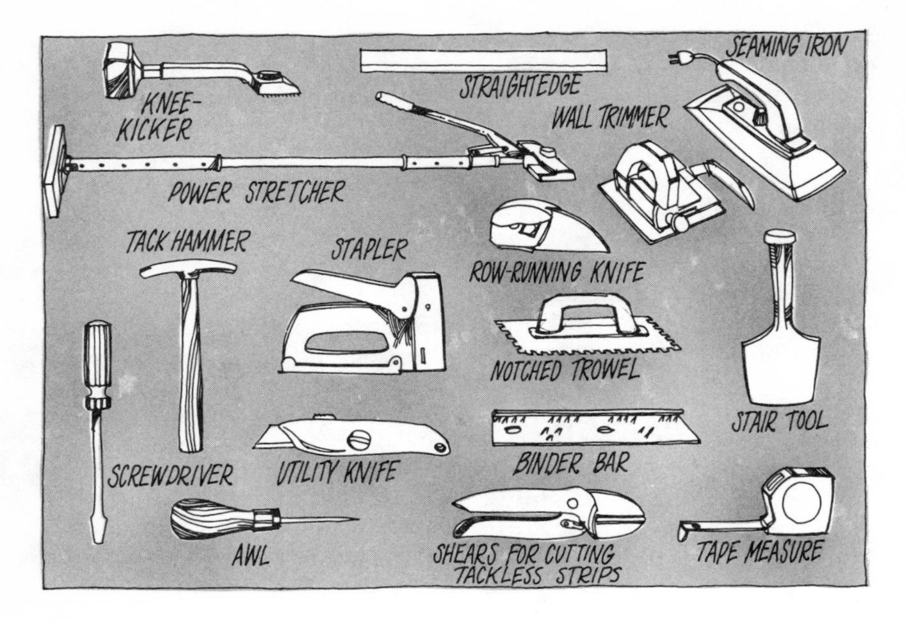

carpet knife, sharp utility knife, linoleum knife, wide-bladed putty knife, and a stapler. If you're carpeting a stairway, you'll also need a carpet awl and stair tool.

To install cushion-backed carpeting with adhesive, you'll need a notched trowel.

Supplies. All the necessary supplies for installing both types of carpeting are available from your carpet dealer.

To secure conventional wall-to-wall carpeting laid under tension, you'll need tackless strips made from ¼-inch lightweight plywood, 1 inch wide and 4 feet long. Each strip is peppered with sharp tacks or pins sticking up at about a 60° angle from the face of the strip. The tacks are 3/16, 7/32, or ¼ inch long.

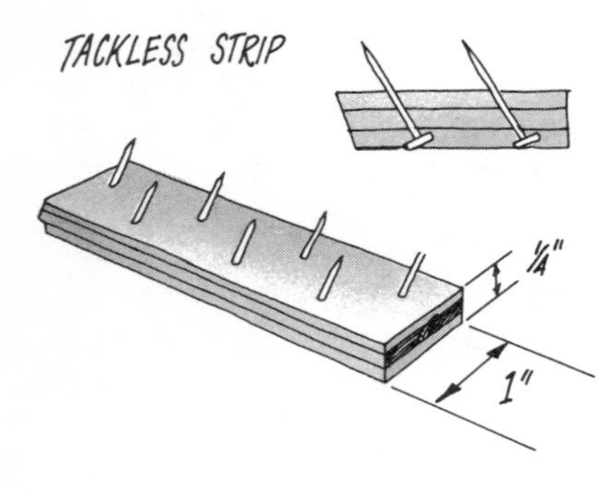

The strips come equipped with nails already in place for attaching to a wood floor, or with masonry nails for use on concrete floors. You can also glue the strips to surfaces that can't be nailed.

To install cushion-backed carpeting, you'll need adhesive or double-faced tape.

Both types of carpeting require binder bars to hold the edges in place in doorways. You may also need tape for piecing padding together, hot-melt or latex tape and adhesive for seaming, and carpet tacks for use on stairways.

PREPARING THE FLOOR OR SUBFLOOR

One of the appeals of both conventional and cushion-backed carpeting is that very little work is necessary to prepare the flooring underneath. Both types of carpeting can be laid over almost any old floor or subfloor, as long as it's clean and reasonably smooth.

Checking the supporting structure. Before installing carpeting, thoroughly examine the floor and supporting structure from below as well as from above. For instructions on what to look for, see page 87.

Making surface repairs. Fix any minor problems—protruding nails, loose boards or tiles, minor squeaks, or an occasional hole or gap. Keep in mind that small indentations and irregularities in the flooring will be absorbed by the carpet padding.

Instructions on repairing floors begin on page 88.

Surface must be dry and clean. Before installing any carpeting, the floor surface must be dry and free from wax, paint, grease, and dirt.

A simple test for checking a concrete slab for moisture is on page 46. Do not carpet slabs that have any moisture content. If you're planning to lay cushion-backed carpeting over dry concrete, first cover the concrete with a sealer.

To remove grease and oil spots, use a chemical garage floor cleaner available at most auto supply stores. Wax or paint should be removed with a floor sander equipped with #4 or #5 open-cut sandpaper (see pages 97 and 99 for tips on renting and operating floor sanders). Finally, sweep or vacuum up all loose particles.

Preparing the room. Remove any shoe molding from around the baseboards and any floor grates or other obstructions. Number the pieces of molding with chalk or pencil if you're planning to replace them.

INSTALLING CONVENTIONAL WALL-TO-WALL CARPETING

Once you're satisfied that the surface of the floor or subfloor is dry, clean, and in as good shape as possible, you can begin the actual work of installing the carpet.

Placing tackless strips and binder bars

Begin by placing lengths of tackless strips in a continuous line around the perimeter of the room; make sure the tacks face the wall. You can either cut the strips to fit now or cut and nail as you work. Use heavy-duty shears to make the cuts in the soft wood.

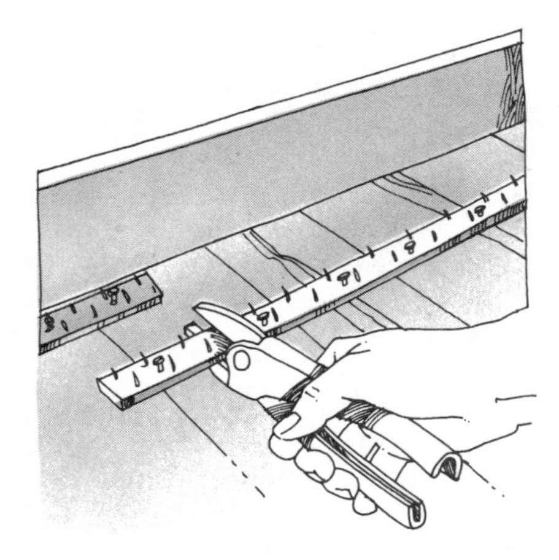

When you're ready to nail the strips in place, position the first one so its distance from the wall is equal to two-thirds the thickness of the carpet. Once you determine the distance, make a guide from a scrap of wood or cardboard (see illustration) so you can position the remaining strips uniformly.

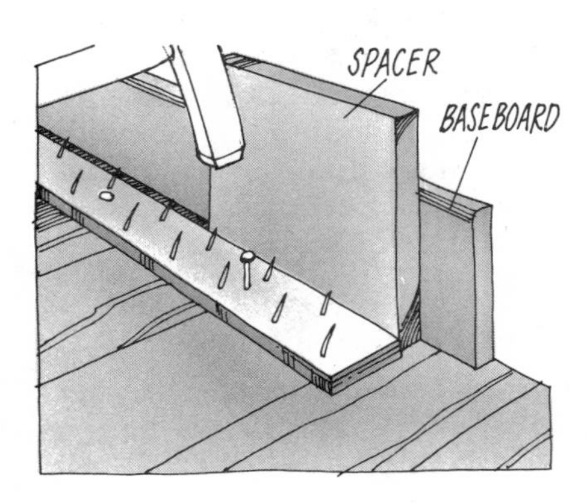

When nailing the strips to the floor, use a tack hammer or a hammer with a small head to avoid damaging the tacks with any misdirected blows.

For driving masonry nails into a concrete subfloor, you'll need a heavier hammer. Be sure to wear safety goggles while driving nails into concrete.

To fasten tackless strips to ceramic tile or any other floor covering that can't be nailed, use contact cement. To get a good bond, follow the directions of the manufacturer and observe all recommended safety precautions.

Attach binder bars in doorways or any other places where the carpeting ends, except against walls (see illustration).

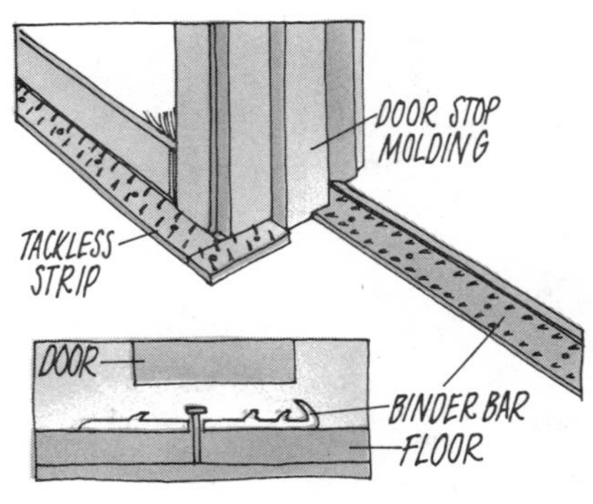

Position each bar so it will be directly under the door when it's closed. After the carpeting is laid, the raised edge of the bar will be folded down over the edge of the carpet. If the door opens away from the carpeted area, you'll have to notch the bar to accommodate the door's stop molding.

Laying the padding

When the tackless strips and binder bars are in place, unroll the padding and cut it so it covers the maximum amount of floor area possible. The padding should overlap the tackless strips just enough to ensure there's adequate padding to fill up the area to the strips; you'll trim it later. If the padding has a wafflelike imprint on one side, place that side up.

If your subfloor is concrete, masonry, or ceramic tile, use the adhesive recommended by your carpet dealer to secure the padding to the surface.

Otherwise, working along the edges of the tackless strips, fasten the padding to the floor with staples placed every 6 to 8 inches.

Cut additional pieces of padding as necessary to cover the entire floor, taping them together and securing the edges with staples. Also staple any areas where you think the padding may slip as you lay the carpet.

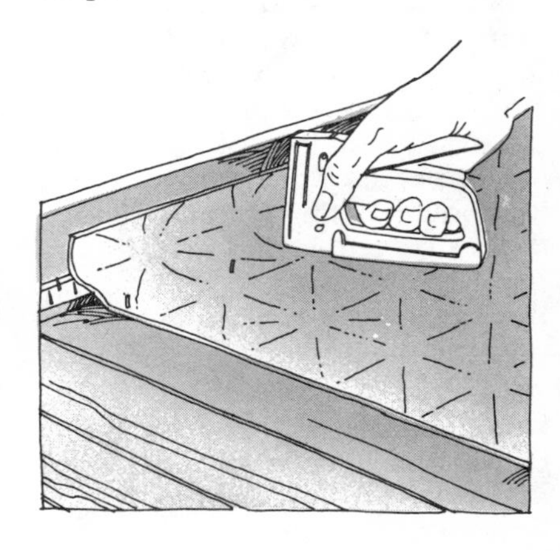

When all the padding is securely in place, take a sharp utility knife and trim off the overlap along the inside edge of the tackless strip, leaving a gap of 1/8 to 1/4 inch between the pad and the strip.

Making the rough cuts

Using the scale drawing you prepared to determine your original carpeting order, roll out the carpet and measure for the cuts you have to make. (If you don't have a large enough area indoors, use your driveway or another clean, flat, and dry area.)

Measure very carefully before making any cuts. Be sure to allow for at least a 3-inch overlap around the edges of the room and for any necessary seams. Make sure that patterns will match and that the pile will run in the same direction.

If you're using cut-pile carpeting, you'll have to cut it from the back. Notch the face of the carpet on both edges to indicate the location of the cut. Then fold the carpet back and snap a chalk line between the notches. Using a straightedge and carpet or utility knife, cut along the chalk line, being careful to cut only through the backing.

Loop-pile carpeting is cut from the front. After snapping a chalk line, use a straightedge and a row-running knife to cut between the rows of loops.

Place the carpeting in the room. If you have to make a seam across the width of the carpeting, place the carpeting to be joined so one piece overlaps the other by 1 inch. Check to see that the top piece is straight. Then take a row-running knife and using the edge of the top piece of carpet as a guide, cut the underlaying carpet. Cut as close to the wall as you can with the row-running knife, then complete the cut with a utility knife.

Seaming the carpeting

A special hot-melt carpet seam tape and seaming iron (both available from your carpeting dealer) will make the most durable and least noticeable seam in synthetic fiber carpeting. Ask the dealer for instructions on their use. You can also use latex tape and adhesive, though it's not as effective as seam tape. Both methods are discussed below.

Using hot-melt seam tape. Take a length of hot-melt seam tape and slip it, adhesive side up, halfway under one carpet edge as shown below.

Holding back one edge of the carpet, slip the preheated seaming iron under the other carpet edge and hold it on the tape, leaving it in place for about 30 seconds. Then draw the iron slowly along the tape as you press the carpet edges down into the heated adhesive. As you work, check to see that the two carpet pieces are butting. If they're not, press them together and place a heavy object on the seam to keep the pieces from pulling apart until they have time to bond to the tape.

Continue working across the floor to where the carpet edges rest against the wall. Pull the carpet away from the wall to complete the seam.

Allow the seam to set before stretching the carpet.

Using latex tape. If the seam's invisibility is not important, you can use latex tape and adhesive instead of hot-melt tape. But the seam will not hold as well under heavy traffic. When applying the tape and adhesive, follow the manufacturer's directions if they differ from the information below.

Cut strips of tape to fit the entire seam. Coat the strips with a thin layer of adhesive. Center the strips under the edges of the carpet.

Squeeze a thin bead of adhesive along the edges of the backing, taking care not to get any adhesive on the pile.

Press the carpet firmly onto the tape, blending the pile with your finger tips. Let the adhesive dry thoroughly before stretching the carpeting.

Stretching the carpet

Once any seams are thoroughly set, you're ready to begin stretching the carpet. Walk around the perimeter of the room and use your feet to shift the carpet so it lies smoothly. Then, with a utility knife, trim away any excessive overlap; you'll still want the carpet to overlap the tackless strips by an inch or two.

Make small relief cuts with a utility knife at each corner of the room so the carpeting will rest flat on the floor. Cut around register grates and other obstacles in the same way, making as many cuts as necessary to allow the carpet to lie flat.

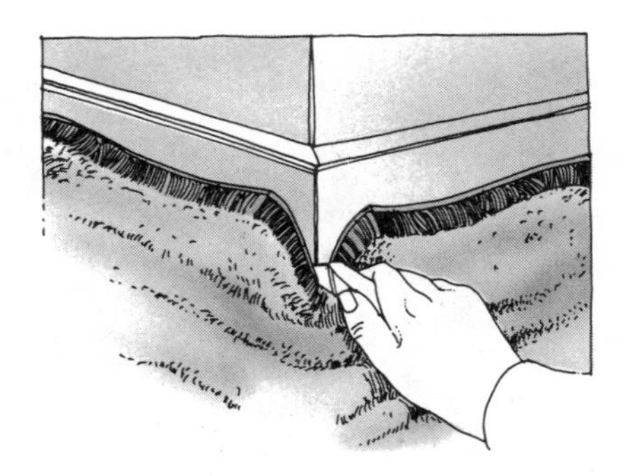

To stretch the carpet, you'll need a knee-kicker to produce the slack necessary to slip the carpet up and over the tackless strips and a power stretcher to pull the carpet across the room so the carpet can be secured on the strips.

Using the knee-kicker. The knee-kicker is a simple device that you can learn to operate with a minimum of practice. Its head has adjustable teeth. Small hooks catch the nap as longer teeth reach down to grip the carpet backing. The opposite end is padded so you can use your knee to bump the knee-kicker forward; the handle telescopes.

To use the knee-kicker, place its head about an inch from the tackless strip. When you bump the pad with your knee, the head will move forward and catch the carpet backing on the pins of the tackless strip, holding the carpeting in place. Use the knee-kicker at a slight angle so the carpeting is stretched along the wall as well as up against the strips.

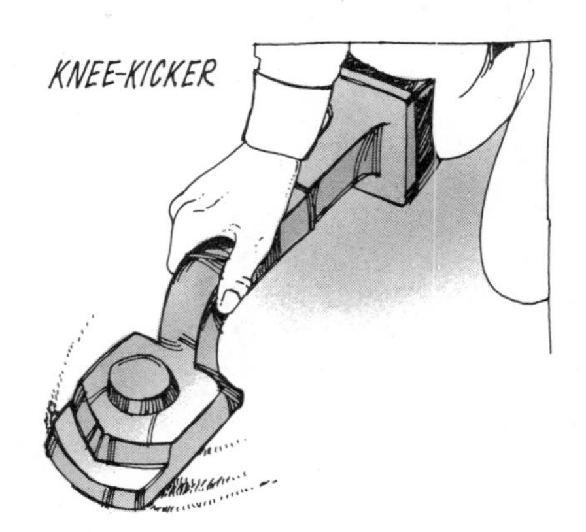

Using the power stretcher. You'll use the power stretcher to pull the carpet across the room toward the opposite wall.

The head of the power stretcher is similar, but larger, than the head of the knee-kicker; adjustable teeth protrude from the head to grip the carpeting. Equipped with a series of handle extensions, the power stretcher can be braced against an opposite wall. When the power stretcher is in place, the "stretch" is achieved by lowering a lever that moves the head forward with steady pressure.

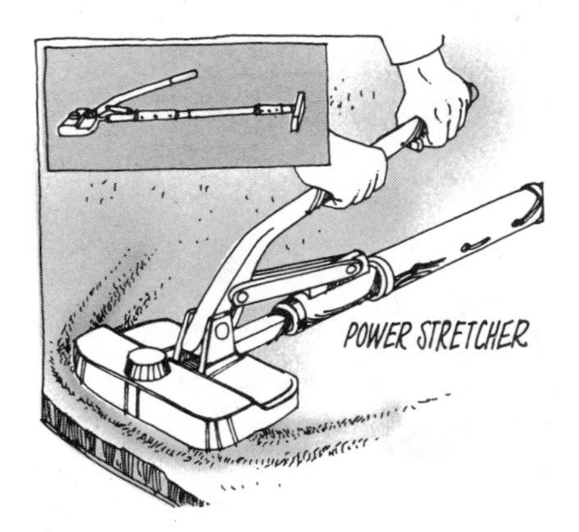

As you work with the power stretcher, experiment to determine how much bite is necessary to get a good grip on the carpet. Stretch the carpet with the minimum force necessary to pull the carpet taut. Too much pressure can tear the carpet.

Protect the baseboards by placing a block of wood cushioned with a piece of carpet between the back of the stretcher and the baseboard.

Securing the carpeting. Follow the sequence outlined below to secure the carpeting to the tackless strips. As you work, consult the illustration below for the proper placements of the knee-kicker and power stretcher.

Step 1: Starting in one corner of the room, use the knee-kicker to secure one corner of the carpeting to the tackless strips along the adjoining walls.

Step 2: Extend the handle of the power stretcher so it extends

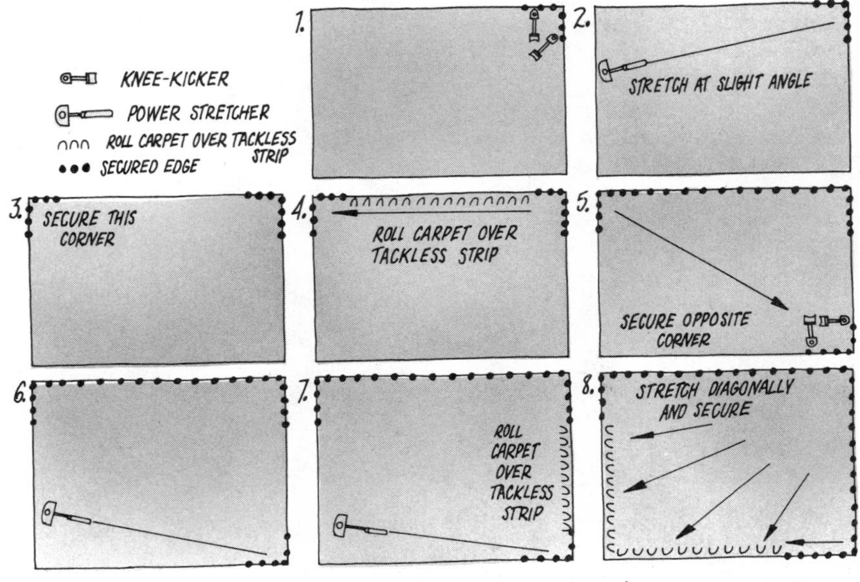

from the corner just secured to the opposite wall (if the room is rectangular, stretch the longer edge of the carpeting first); angle the stretcher slightly so it's not pointed directly into the corner. Bite into the carpeting and secure it to the tackless strip.

Step 3: With your hands or a linoleum knife, roll the corner just stretched to secure it to the tackless strip.

Step 4: Moving along the wall from the corner where you started, use your hands or a linoleum knife to ease the carpeting over the tackless strip.

Step 5: Move down the wall from the corner where you started (the short wall, if the room is rectangular); use the knee-kicker as in Step 1 to secure the corner of the carpeting to the tackless strips along the adjoining walls.

Step 6: With the power stretcher at a slight angle (as in Step 2), secure the carpet to the opposite wall a short distance from the corner.

Step 7: Roll the carpeting over the tackless strip (as in Step 4) along the wall between the two secured corners.

Step 8: Fanning out diagonally from the first corner you secured, use the power stretcher to stretch the carpet over the tackless strips along the two walls not yet secured. With your hands or a linoleum knife, roll the carpet over the strips to hold the carpeting in place.

Finishing the job

Trimming the carpeting between the wall and the tackless strip is the crucial last step for installation. If possible, use a wall trimmer designed specifically for making the final cut around the perimeter of a room; otherwise, use a utility knife for all cuts.

Adjust the trimmer for the thickness of the carpet. Leaving just enough carpet to be folded over the tackless strip and down into the gap between the strip and the wall, slice downward into the carpeting.

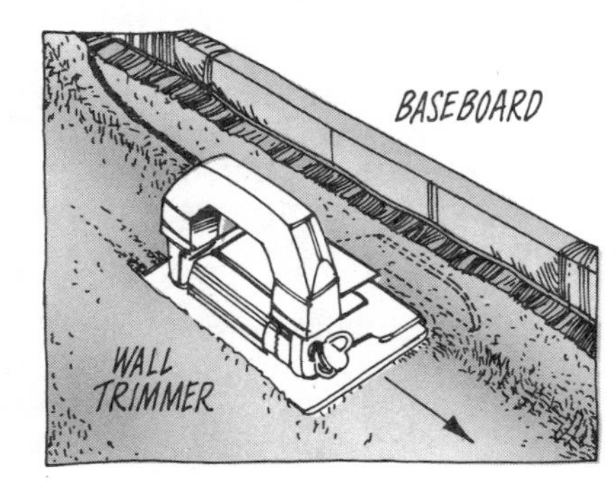

In corners and around obstacles, use a utility knife to make the cuts.

Using a trowel or wide-bladed putty knife, force the edge of the carpeting down into the gap between the tackless strip and the wall. If the carpet bunches up, trim away additional carpeting as necessary.

Where you've used binder bars, trim the carpet to fit; then take a block of wood and a hammer and rap the top edge of the binder bar to bend it tightly over the edge of the carpet.

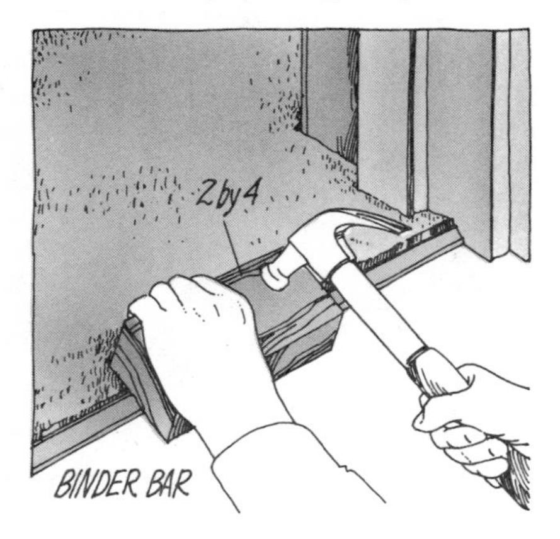

Though most wall-to-wall carpeting is installed without shoe molding, you may decide to use it. Examine the job after you're done and replace the shoe molding, if you feel it adds to the appearance.

INSTALLING CUSHION-BACKED CARPETING

Cushion-backed carpeting is easier to install than conventional wall-to-wall carpeting, since it doesn't require stretching and its pad is bonded in place. Secure cushion-backed carpeting to the floor either with adhesive or with double-faced tape.

Adhesive has the advantage of providing a firm bond. But once the carpet is set, it can't be lifted without destroying the rubber backing and leaving adhesive and scraps of backing on the flooring.

Double-faced tape is easy to use, but it doesn't work well on seams and doesn't hold the carpeting securely.

Preparation of the subfloor for cushion-backed carpeting is exactly the same as it is for conventional carpeting (see page 78).

Making the rough cut

Roll out the carpeting in a large area that's clean, dry, and flat. Then, using the scale drawing you prepared for ordering the carpet, measure for the initial rough cut. Allow for at least a 2-inch overlap around the edges of the room and for any necessary seams. You'll do any required seaming after the rough cut is made and the carpet sections are placed in position in the room.

Double-check your measurements before making any cuts. Also check to make sure that patterns will match and that the pile will run in the same direction. Except that cushion-backed carpeting is cut right side up, cut the carpeting as you would conventional carpeting (see page 80).

Fitting the carpet

If you're planning to use adhesive and need to make a seam, use the instructions that follow. Directions for using double-faced tape are at right, below.

Using adhesive and making seams. Move the carpeting into position in the room. If you need to make a seam, measure across the floor to where the seam will fall; then snap a chalk line. Line up a piece of carpeting so the carpet edge falls on the chalk line.

Position the second piece of carpeting so it overlaps the first by at least ¼ inch. Fold both pieces back to expose about 2 feet of the floor on both sides of the chalk line. Using the notched trowel, spread adhesive evenly over the exposed floor area.

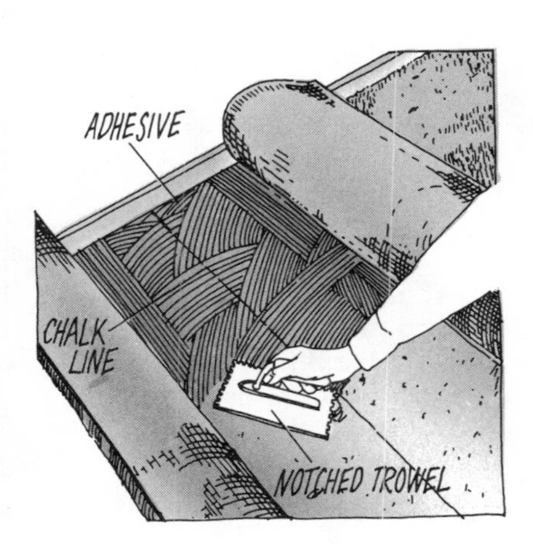

Roll the edge of one carpet section onto the adhesive, keeping the edge aligned with the chalk line. Following the directions provided by the seam adhesive manufacturer, apply a bead of seam adhesive to the backing between the cushion and the carpet pile the full length of the seam (see illustration).

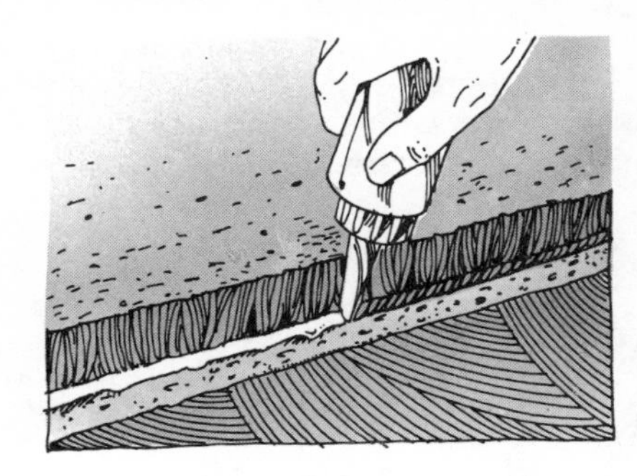

Unroll the other carpet piece and butt it against the edge of carpeting already in place. Use your hands to press the carpet into place; work away from the seam to force out the bubbles of trapped air.

When the seam adhesive has thoroughly dried, take hold of one loose end of carpeting and pull it back over the seam to expose the floor that hasn't been covered with adhesive.

Using the trowel, finish applying adhesive. Then carefully roll the carpeting onto the adhesive, working out air bubbles as you go. Repeat the same steps to secure the carpeting on the other side of the seam.

When the carpet is firmly set in the adhesive, you're ready to trim off the excess carpeting. Working around the perimeter of the room, use a utility knife to trim the carpet along the baseboards, leaving carpet equal to about one thickness lapped up the wall. Use a trowel or wide-bladed putty knife to tuck the overlap down against the wall.

Use binder bars to cover carpet edges in doorways (see page 79).

The tape alternative. Double-faced tape is a fast and inexpensive method of holding cushion-backed carpeting in place. But it's not recommended for use in rooms with heavy traffic and isn't suitable for securing carpeting that's seamed.

If permanence isn't a major consideration, double-faced tape will do an adequate job. Working your way around the edges of the room, fasten tape to the floor (leave the top protective paper covering in place). Then simply fit the carpet into the room.

If it's convenient, leave the carpet laying loose for 2 or 3 days to flatten it out. Then remove the protective covering from the tape and press the carpet edges into place over the tape. Use a utility knife to trim the carpet to fit.

INSTALLING CARPETING ON STAIRS

Many homeowners with wall-to-wall carpeting prefer to extend carpeting up stairways for esthetic reasons. But there are practical reasons for carpeting stairs as well. Carpeted stairs are quieter and safer, reducing the likelihood of falls and providing a cushioned surface should such a mishap occur.

Because carpet on stairs receives heavy wear, conventional carpeting of good quality with durable padding is recommended. Cushion-backed carpeting is not likely to give satisfactory service.

Measuring stairways

A number of variables make measuring stairways for carpeting more complicated than measuring a standard room. The steps themselves may be irregular. In addition, allowances must be made for landings and for odd-shaped steps where stairs wind around corners.

Enclosed stairways are usually carpeted wall-to-wall. Stairways with one open side are generally covered from the wall to the balusters on the open side. In both cases, stairway carpeting is rolled under about 1¼ inches on both sides, another consideration when ordering.

To determine how much carpeting you need, measure the distance from the back of one tread to the bottom of the riser below as shown in the illustration. Add another inch to that total to allow for padding under the carpet. Multiply that figure by the number of steps to get the total

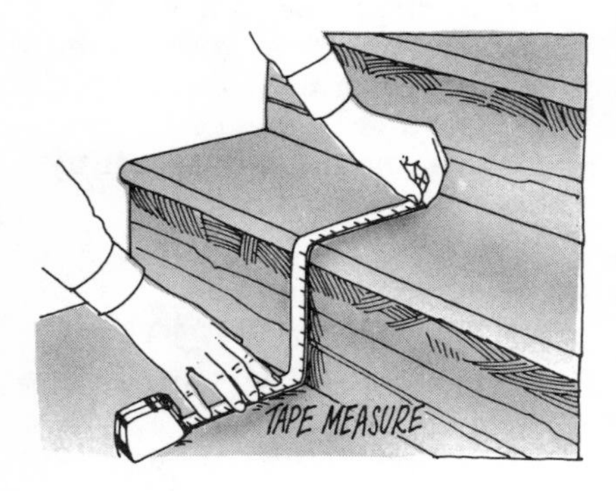

length required. Measure the width from wall-to-wall or from wall to balusters (remember to add a total of 2½ inches for the folded-under edges). Also measure for any landing at the top of the stairs.

Take the measurements with you to your carpeting supplier, who can recommend how much you need to buy and how you can make the most economical cuts.

Installing tackless strips

The same tackless strips that hold conventional wall-to-wall carpeting in place are used to hold carpeting in place on stairways. You'll install tackless strips at the back of each tread and the bottom of each riser.

Cut each piece of tackless strip 2½ inches shorter than the width you're covering (to allow for the folded-under borders). Cut a scrap of wood ⅝ inch thick to use as a spacer. Working down the stairway, rest the strip on the spacer and nail the first piece of tackless strip ⅝ inch up from the tread; be sure that it's centered on the face of the riser and that the pins point down.

Nail a second strip (also centered) near the back of the top tread ⅝ inch out from the riser (pins should point toward riser). Continue adding strips down the stairs as illustrated at the top of the column on the next page. If your stairway has a landing, carpet it in the same way wall-to-wall carpeting is installed in a room; leave enough of an overhang on the down side of the

CARING FOR CARPETING

Regardless of the fiber used, proper care of carpeting requires regular cleaning, the best form of preventive maintenance. Invisible dirt particles on the surface—not foot traffic—do the damage that wears out carpeting. Grit that's not removed chews into the fabric as the carpet is walked on.

Normally, vacuuming carpeting once or twice a week should be sufficient for removing dirt. The vacuuming should be done whether the surface looks dirty or not. Always use a vacuum cleaner that's in good condition; one with a vibrating beater bar is considered best for all-round use.

Occasionally, an overall cleaning is necessary to get rid of accumulated dirt and restore a carpet to its original brightness. The three basic methods you can use to do a thorough cleaning are hot-water extraction, wet shampoo, and dry cleaning. Each method has its advantages and disadvantages.

Hot-water extraction. Because it doesn't employ mechanical brushing that can distort pile and cause flaring of yarn tufts, hot-water extraction (sometimes called steam cleaning) is generally recommended as the best method for cleaning most types of carpeting.

The system forces hot water into the carpet, then vacuums the water and dirt solution out of the carpeting. A strong vacuum unit is essential in this system, since too much water left in the carpet can cause the backing to shrink, mildew, or stain.

Wet shampoo. This system is the one used most often by do-it-yourselfers because the cleaning units are widely available for rent. A rotary brush works a detergent foam into the carpeting to loosen dirt particles. The dirt is suspended in the foam while the solution dries. When it's dry, the carpet is vacuumed and the residue removed.

When shampooing a carpet, it's best to use the minimum possible amount of detergent to avoid leaving detergent residue in the carpet. Any residue causes the carpeting to get soiled faster and leaves a dull finish. If you shampoo your carpets regularly (about once a year), change to a hot-water extraction method every two or three cleanings.

Dry cleaning. Powdered compounds that can be worked down into the pile of a carpet with no special equipment are available. You simply brush the dry compound onto the carpet and work it in, then go over the carpet with a vacuum cleaner.

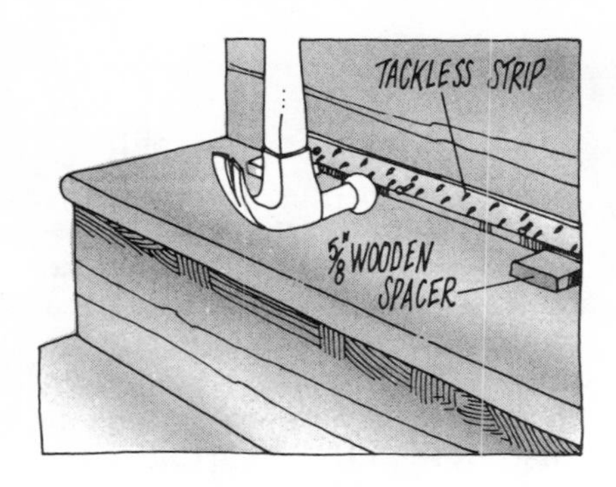

landing so the carpeting can be wrapped down to cover the riser of the step below.

Installing padding

Individual pieces of padding must be cut and stapled to each step. Cut each piece of padding the same width as the tackless strips. Butt the padding against the strip at the back of the tread; then fold it forward over the edge of the tread, cutting it about 3 inches above the tackless strip on the riser below. Staple the padding to the stairs along the top and bottom edges as shown.

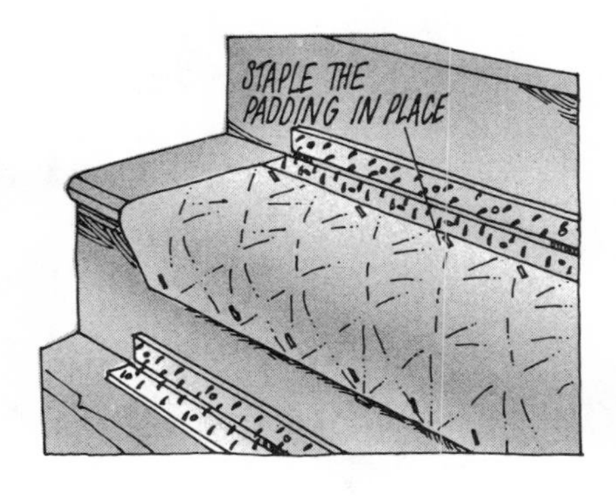

Securing the carpeting

Start at the bottom of the stairway to install the carpeting, but place it so the pile runs downward. Carpet laid with the pile running uphill will wear about half as long.

Preparing the edges. Cut the carpeting into a strip or strips of the proper width. Then place each strip face down on a flat working area. Take a chalk line and snap a line along each edge of the carpeting 1¼ inches from the edge. Use an awl and a straightedge to score the back of the carpet as shown below. As it's scored, fold the edges over.

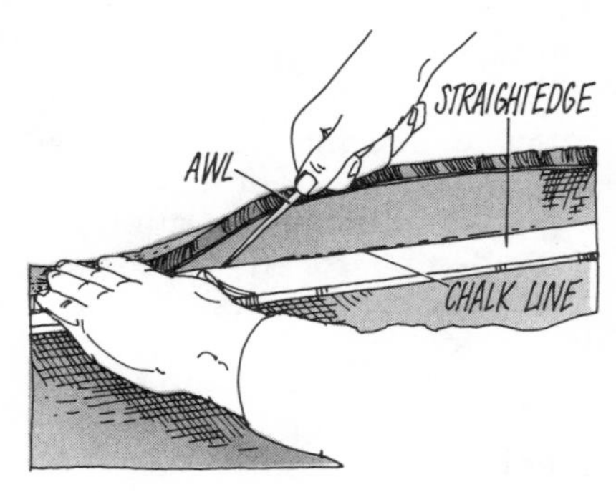

When the edges have been folded the full length of each strip, place the carpet into position on the stairs.

Begin at the bottom step. Center the carpet on the stairs with the lower edge overlapping the floor about one carpet thickness and the folded-under edges falling outside the ends of the tackless strips. Use an awl to ease the carpet down over the pins of the tackless strip on the bottom of the riser. Press the carpet onto the pins. Then drive a carpet tack through the folded-under edge on each side and into the riser. Tuck the bottom edge of the carpet into the gap between the tackless strip and the floor.

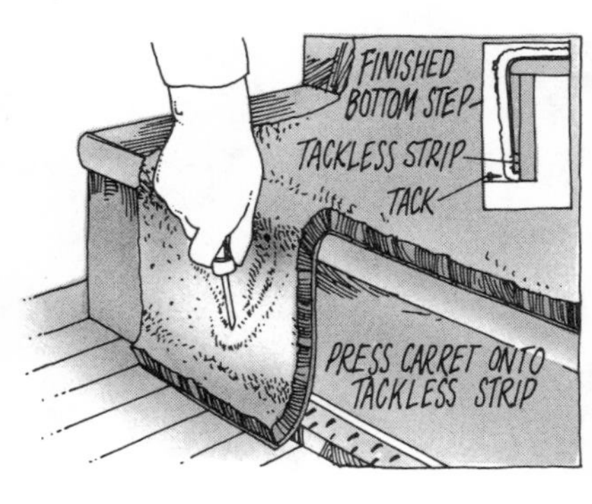

Using the knee-kicker. When the lower edge of the carpet is secure, place the head of the knee-kicker in the center of the tread and bump the carpet taut. At the same time, use a stair tool to force the carpeting down into the space between the two tackless strips as shown. Once the carpet is firmly secured in the center, angle the knee-kicker toward each side of the step and move across the step securing the carpeting.

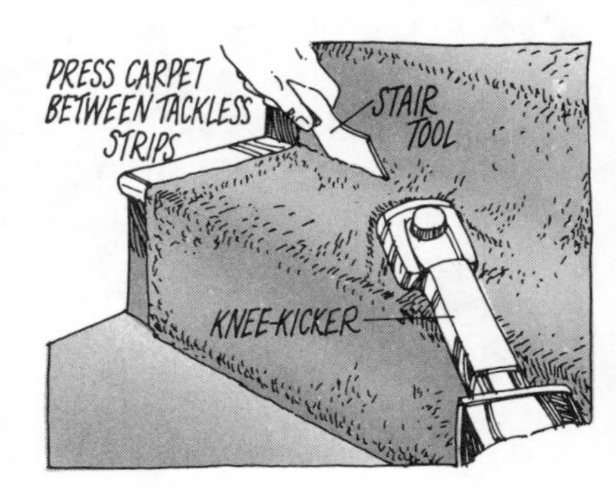

Then drive a carpet tack through the folded edge on each side of the step into the corner where the tread and riser meet as illustrated.

Stretch and secure the carpeting one step at a time. If you need to join two strips of carpeting, cut the end of one piece so it will fit over the tackless strip on the tread and tuck into the space behind it. Begin the new section of carpeting so the lower end tucks into the same gap. Make sure the pile continues to run down the stairs.

If the carpeting ends at the top of the stairs, use a binder bar to hide the exposed end. Don't end the carpet directly on top of the final step.

REPAIRING & REFINISHING FLOORS

A handy guide to most flooring repairs—from simple to not-so-simple

- **sagging floors—their causes and remedies**
- **squeaking floors—annoying but fixable**
- **squeaking stairs—several solutions**
- **refinishing wood floors—exacting but rewarding**
- **floor surfaces—repairs that work wonders**
- **carpet repairs—tricky but not impossible**

Whether planning minor repairs to the surface of a damaged floor or considering a major renovation, the prudent homeowner will make sure that the floor is in good structural condition before beginning any finishing work.

Surface problems such as cracked ceramic tiles, loosened or cracked resilient flooring, or squeaks in floors and subfloors can often be traced to defects in the subfloor, joists, girders, posts, footings, or even foundation sills and wall studs (see drawing, facing page). Fortunately, some common structural flaws are relatively simple to correct with readily available tools.

Structural problems caused by stress or deterioration in the foundation require the special talents of a contractor. Rot or decay, the working of time and moisture, can make it necessary to reinforce or replace weakened joists, girders, or posts. Sills and studs are also susceptible to decay and moisture damage.

Reinforcing weak joists, replacing posts, or using adjustable jack posts to support sagging joists are projects that can be undertaken by the home handyperson. Replacing girders or joists or repairing sills and studs are jobs best left to professionals.

Examining the underside of your floors may not be as rewarding as getting on with the job of bringing an old oak floor back to life. But resist the urge to skip the preliminaries. A little time spent checking in the basement or crawl space under the house may save you a lot of time, expense, and frustration in the future. If your inspection shows that everything is in good order, you can go ahead with your flooring project confident that you're starting with a solid base.

SAGGING FLOORS—THEIR CAUSES AND REMEDIES

A few minor dips in floors, particularly in older homes, are not uncommon. Over a period of years, some settling is bound to result from gradually applied stresses and fatigue. But if floors sag or are noticeably springy, the problem could range in scope from modest to serious.

Consider the gravest possibility first. If there are cracks in the plaster or other wall materials in or near the house's normal stress points—in corners, over windows and doors, or where walls meet—more than simple structural repairs may be necessary.

Damage beyond the floors may be a sign of a deteriorating foundation or excessive settling. For example, horizontal movement and the cracking of foundations sometimes occurs in houses located on sloping ground or near cut slopes, often as a result of inadequate diversion of surface or roof water.

Sticking doors or windows or leaky plumbing may also be signs of foundation problems.

Many older homes may have uneven floors as the result of settling that occurred shortly after the house was built, even though the house was well built and may have been stable for many years. Before deciding to correct a problem of such long standing, consult an engineer or builder familiar with older homes. The results may not be worth the risks involved.

Solving major problems may involve the complex and expensive job of raising the house to permit structural and foundation repairs—clearly a job for a contractor with special skills and equipment. Happily, trouble in the foundation is one of the least common causes of sagging floors.

Locating the cause

The first thing you must do is isolate the cause of sagging floors and determine where weaknesses exist. On the following pages, you'll find a logical series of checks you can make. The problem may be caused by a single defect, like a settled footing or a weak splice in a beam, or it may be the result of a combination of two or more weaknesses.

Working from above, you'll first have to locate low spots; then from underneath the floor, you'll have to make a detailed examination of joists, girders, posts, footings, foundation sills, and wall studs where they rest on sills. In most older homes, girders and posts, like joists, are made of wood and are vulnerable to deterioration. In newer homes, steel girders and support posts are common. These are very unlikely to break down, but may direct your attention to problems in the foundation or footings.

In any case, it's important that you make a *thorough* inspection of the supporting structure beneath a sagging floor. If you stop, say, at the joists just because you think you've found the cause of the problem there, defects in the girders, posts, or footings may go undetected—and come back to haunt you later.

If you're unable to pinpoint the cause of a sagging floor, you'll need the advice and assistance of a professional.

Find the low spot. Take a straightedge at least 8 feet long and check systematically for the low spot in the floor. A straightedge can be a length of rigid pipe; a piece of lumber at least 4 inches wide, selected for straightness;

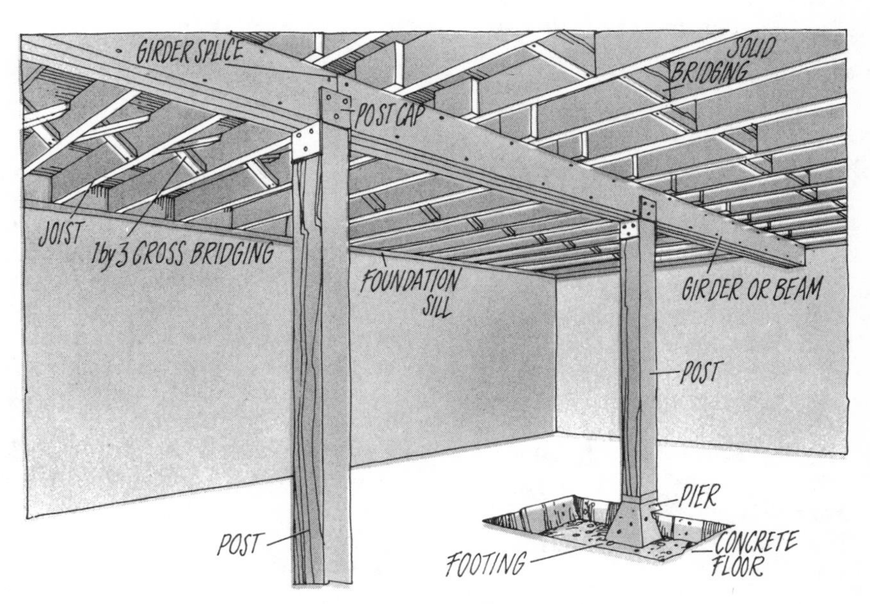

Supporting structure for floor in typical frame house consists of joists, girders, and posts resting on concrete piers and footings. Bridging adds strength to joists.

or the uncut edge of a strip of ¾-inch plywood.

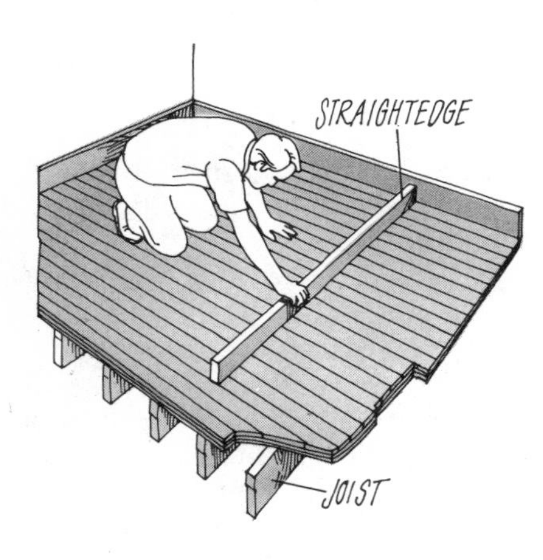

Once you find the low area, mark the spot. Then take a tape measure and "map" the spot in relation to two or more reference points (corners, ducts, pipes) that you can use to locate it on the underside of the floor from the basement or crawl space.

Checking joists and girders. Here's what to look for. Joists should be checked for sway, sagging, or warping. Use an awl to poke into the wood, particularly where joists rest on foundation sills or adjoin other wood members, and look for rot or insect damage. Extreme softness in the wood is evidence of rot. If termites have been feasting on your home, you'll discover their tunnels in the wood or on the foundation, and you'll need immediate help from an exterminator.

Discolored wood, usually dark, on joists or in the subfloor can indicate a current or recent plumbing leak or exterior water leaks. Moisture can also gather where there is inadequate ventilation below floor level. Moisture attracts subterranean termites and promotes fungus growth, which ultimately results in rot.

If the joists are straight and appear to be sound, examine the girders. Check carefully where the girders rest in pockets in the foundation walls or on wood sills. And if the girders are made up of varying lengths of wood, check the points at which they're joined—weaknesses commonly occur where splices have been made other than directly over supporting posts.

Checking posts, piers, and footings. If the low spot in a sagging floor is over or close to a post and the joists and beams are in good shape, the post or its pier or footing may be in bad condition.

Check wood posts for signs of deterioration, especially where they sit on a pier or a footing; moisture here can cause decay. Inspect wood posts thoroughly for termite damage too.

If there's no apparent decay or insect damage, or if your posts are made of steel, then the footing has cracked or settled, allowing the whole structure above to sink. The best solution is to replace the footing and, if necessary, the post (see "Is the footing adequate?" on page 91).

Checking the concrete slab. If your home is built on a concrete slab and the slab has cracked substantially or tilted, it usually means the slab was poured on insufficiently compacted fill or on unstable land. Get professional advice.

Hairline or narrow cracks in exposed concrete slab floors are usually not serious and are probably unrelated to structural problems. Once discovered, they can be patched easily with one of several commercial products made for the purpose and available from a hardware dealer or home improvement center. If you're unsure how serious a crack is, check with a professional.

Upper floor sags

Multistory homes may have problems in the upper floors identical to those on the first floor because they're directly related.

For example, a load-bearing wall that has settled or weakened may cause both the first and second floors to sag because both are supported by the same structure. In this case, leveling the first floor may level the second floor as well (see "You must level the floor," facing page).

If an upper floor sags between the exterior wall and an interior load-bearing wall, correcting the problem can be complex. Have a structural engineer or capable contractor examine the structure. Because the underside of the floor is probably the finished ceiling of the room below, repairs can be messy and costly—not a job for the homeowner. If there are no serious structural complications, it may be easiest to live with the slope in the floor.

Should you tackle the job yourself?

Once your preliminary investigations have turned up the specific weaknesses that have allowed your floor to sag, the next step is to decide if you want to tackle the necessary repairs yourself. Reading through the repair instructions in the next few pages will help you understand what's involved.

But before you decide to take on the job, especially if you're uncertain about what might be involved, consult a structural engineer familiar with local building codes. The engineer can help you determine which repairs are practical for you to do yourself, and can provide specifications and a list of materials that will be needed. A building inspector should also be contacted early in the planning phase so that building permits can be acquired if necessary.

If the general structure supporting the floor is in good condition and the sagging is minor—

say, less than ¼ inch over an 8-foot span—you may choose to live with the imperfection. You can correct minor dips by driving wooden shims between the joists and subfloor, much as shims are used to eliminate squeaks or to stabilize springy floors (see "Getting rid of the squeaks," page 93).

The most common structural repairs that the homeowner may elect to do are doubling or reinforcing joists, replacing posts or footings, and reinforcing girders. Reinforcing a joist, however, can be difficult if it involves working around or disturbing wiring, plumbing, or duct work.

Replacing beams and joists or repairing sills and studs are jobs best left to professionals; the work can be difficult, and it requires skill and capable assistants. Only the most talented do-it-yourselfer should attempt major structural repairs—and then only in straightforward situations when professional advice is available.

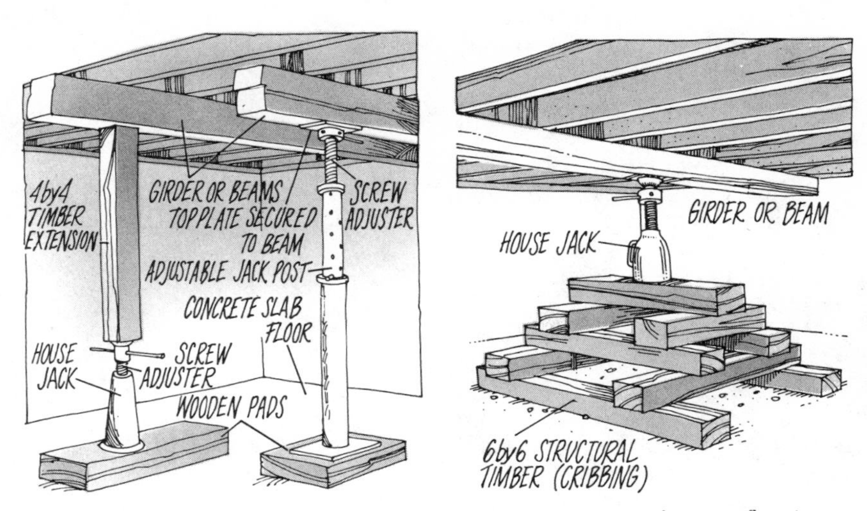

Basic jacks—either house jacks or adjustable jack posts—are used to raise floor in frame house. Adjustable jack posts can be left in place as permanent support posts.

You must level the floor. Correcting any sagging floor means first bringing the floor back to level. In most cases, this will involve some heavy work with heavy pieces of equipment. Structural timbers will be needed as well as one or more jacks.

Understanding the basic jacks. House jacks are required for the basic business of raising a structure while repairs are made. You can also raise the structure with jack posts and leave them in place as permanent support for a newly leveled floor.

For houses with basements, either a basic house jack with a timber extension or an adjustable jack post may be used. If your house has crawl space only, you'll need a house jack set on a pyramid or cribbing built of structural timbers, usually 6 by 6s. The pyramid provides a solid foundation for the jack and raises it high enough to reach the girders or joists.

The adjustable jack post is fitted top and bottom with heavy metal plates that can be secured temporarily to a wooden pad and a beam with duplex nails—double-headed nails that are easy to remove.

The jacks should be "screw-type" jacks—not hydraulic, which are difficult to adjust in the fine gradations required for raising a house slowly enough to avoid damaging it.

Any rental company that supplies jacks should have structural timbers available or be able to direct you to a source. Railroad ties or other used timbers are often available from house movers or demolition companies.

How to double a joist

The easiest way to strengthen a weakened or decaying joist is to "double" it by securing a new joist to the old. The new joist should be the same thickness and width as the old one, but need not be long enough to sit on the foundation sill or on a girder.

Before you double a joist, you should have an exterminator treat any decay or insect damage on or around the old joist.

Doubling of joists should be done only in moisture-free areas. Dampness seeping between the two joists can cause rot that will be impossible to detect on their surfaces.

Placing the jack. To begin, remove any bridging between the sagging joists (see drawing). Set

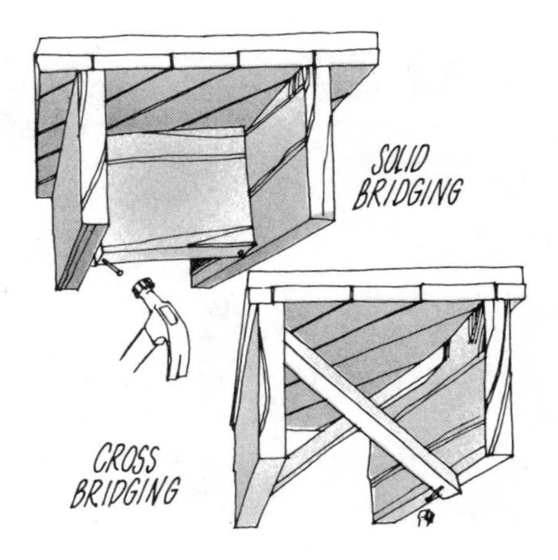

a house jack on the top timber of the cribbing, or set a jack post on a wooden pad made out of a 4 by 8 or wider timber. You can also

use a house jack, set on a wooden pad, with a 4 by 4 timber extension. Use the last two methods only where a concrete slab or footing is available as a base (see "Understanding the basic jacks," page 89).

Place a 4 by 6 or 4 by 8 beam, long enough to span the area of the floor to be raised (a minimum of 4 feet), perpendicular to the joists. Set 2 by 6 blocks on top of the beam so they will be directly under each joist when the beam is raised into place. This will make it easier to slide the reinforcing joist into place once the floor above is level.

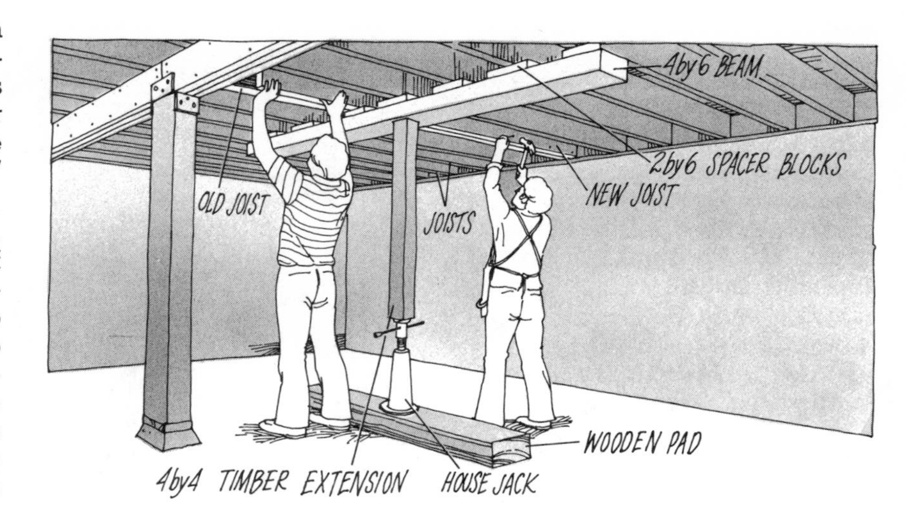

Extra hands will be required to raise the beam and hold it in place while the jack is made plumb and enough pressure is applied to hold the beam snugly against the joists (see drawing at right).

If you're using an adjustable jack post, nail the bottom plate to the wooden pad and the top plate to the beam.

Raising the floor. Grease the threads of the jack and start raising the floor 1/16 inch every 24 hours. Because the threads vary among different types of jacks, you'll have to determine how much of a turn is required for the jack you're using.

Depending on how high the floor needs to be raised to eliminate the sag, the process can take several days. To allow for some settling, the floor should be raised to 1/4 inch above level.

Resist the temptation to accelerate the process. The house will need time to make subtle adjustments to the new stresses created as the floor is raised. If you try to hurry it, you'll risk cracked plaster and structural damage.

Installing the joist. Coat the surfaces to be joined with a waterproof wood glue; then lift the new joist into place and secure it with nails or bolts.

If nailing is your preference, use 16-penny common nails staggered top and bottom about every foot, and clinch the protruding points by bending them flush with the surface.

With the limited space between joists to swing a hammer, 1/2-inch bolts may be easier to install and will work as well as nails. Predrill bolt holes in the new joist about every 18 inches before lifting it into place. Secure it with a few nails or C-clamps, then use the holes in the new joist as guides to drill through the old joist. Insert the bolts and tighten.

Once the new joists have been added, lower the jack at the same rate of 1/16 inch per day until it can be removed.

Install new bridging between the joists, and the job is finished. You can choose between diagonal bridging or solid block bridging.

Diagonal bridging can be cut from 1 by 3s or 1 by 4s. One end of each piece is nailed to the upper edge of one joist and the other end to the bottom edge of the adjacent joist (see drawing, page 89).

Solid block bridging can be cut from lumber the same size as the joists to fit at right angles between joists; the pieces are then nailed in place. Solid block bridging is usually staggered to make it possible to nail through the joists and secure the bridging from both ends (see drawing, page 89).

How to replace a post and footing

Replacing a post is one of the less complicated structural repairs. A homeowner can tackle the job with one or two helpers, some professional counsel, and the necessary local building permits. Before buying a new post, check with your building department to determine what size you'll need.

The easiest post to install is one made of wood the same size (not length) as the post you're replacing. Wood pressure-treated for termite and rot resistance is best. Using a saw, you can cut a wood post to the exact length required.

A steel post, on the other hand, must be made by a steel fabricator to the exact length required; while you can shim it if it's too short, you'll have no way of cutting it if it's too long. A steel post comes with predrilled flanges top and bottom.

Another alternative is to use an adjustable jack post, but you should check first with the local building department.

Raising the structure. To make it possible to remove an old post, use adjustable jack posts or house jacks to level the floor and take the weight off the post. Place two jacks under the girder, 3 feet from each side of the post (see drawing below), and make sure they're plumb. Raise the floor 1/16 inch per day until it's 1/4 inch above level. Patience is required, as the structure may be damaged if an attempt is made to hurry this step.

Removing the old post. While raising the structure, remove any visible nails or bolts that fasten the post to the beam. By the time the floor is level, you should be able to remove the post. If the post doesn't slide away, it may be held by a pin at the bottom or there may still be a load on it.

If you're sure the weight has been lifted free and the post still proves stubborn, take a saw and cut through the post near the top at a slight angle. The lower portion should come away easily. Then remove the top portion if it's still attached.

Is the footing adequate? If the post beneath a sagging floor is in good condition, then the footing has probably cracked or settled. Even when you've identified a deteriorated post as the apparent cause of sagging, check the footing on which it sits. Though not cracked or settled, it may be too small for the job it's being asked to do. Ask a building inspector for the correct size—footings 2 feet square and 2 feet thick are common.

Removing an old footing—and the concrete slab that may be covering it—can be heavy work. There's no easy way of doing it. You can rent an electric jackhammer to do the job, or you can break up the concrete into chunks with a sledge. Or you may want to hire a professional.

Replacing a footing. To prepare for a new footing, dig a hole the required size and dampen the soil. If your building inspector requires the top of the new footing to project above the adjacent slab level (many do), you'll need to build a form to contain the concrete to be poured above the slab. No form is needed below the level of the slab as the earth walls of the hole provide a natural form.

To determine the amount of concrete you'll need for the footing, calculate the volume of the footing in cubic feet; don't forget to add in the amount you'll need to repair the slab around the footing. Add the volumes together if you have more than one footing to pour. Adjust the recipe that follows for the amount of concrete you'll need.

To make 10 cubic feet of 1:2:3 concrete, you'll need 1⅘ sacks of cement, 4½ cubic feet of sand, 9 cubic feet of ¾-inch gravel, and 9 gallons of water.

The amount of water is based on 5 gallons per 94-pound sack of cement, and sand of average wetness. If your sand is very wet, use about 4¼ gallons of water per sack of cement. For barely damp sand, increase the water to 5½ gallons per sack.

For small quantities of concrete, you can purchase 90-pound sacks containing a dry mix of cement, sand, and gravel. Each sack will make about ⅔ cubic foot of finished concrete.

When the concrete needed is about ⅓ yard, as it is for a 2-foot cube, it can be mixed easily by hand. A wheelbarrow is convenient for mixing; each batch can be poured directly into the hole (see drawing). Vibrating or stirring the concrete in place will help eliminate air pockets.

Use a straight board to scrape off any excess concrete above the level of the form. If you're going to install a wood post on the new footing, a stock metal post base should be set in the concrete while it's still wet. You can also use a precast pier (with a wood nailing insert) set on the wet concrete (see drawing above). If you're going to install a steel post, it will have a flange with predrilled holes for anchor bolts. The bolts should be set in the concrete before it sets.

When the base or pier is in place, cover the fresh concrete with a sheet of polyethylene and keep the concrete moist for 2 weeks until it cures. Let it cure thoroughly before installing a new post on it.

Installing a new post. Once your new footing has cured, or you've made sure that an old footing is not defective, you're ready to install the new post. Wood posts and steel posts are installed in slightly different ways; both techniques are described in the following paragraphs.

But first, check to see if the beam above is spliced directly over the post you'll be installing. If it is, you'll want to reinforce the beam before setting the post.

Unless you're using a wood post (in which case the prefabricated post cap you'll use to install the post will do the job of reinforcing the beam), you'll have

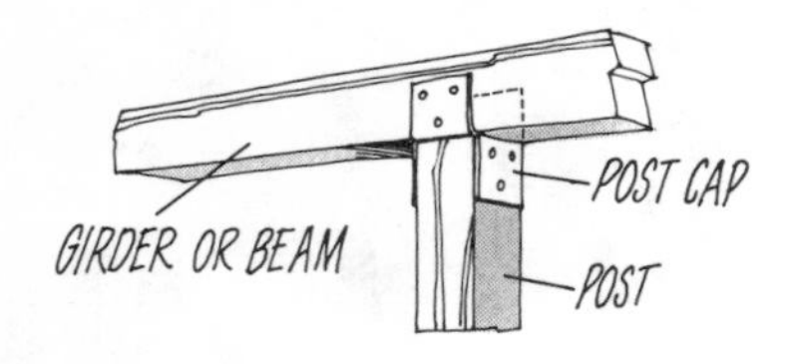

to cut two pieces of ¾-inch plywood as wide as the beam and 4 feet long. Center the pieces over the splice, one on either side, and fasten them to the beam with 8-penny common nails or ¼-inch lag bolts.

Now you can set the post in place. If it's a wood post, put it on the post base or pier and nail it to the base or insert, but do not drive the nails home. Then fit the post cap, plumb the post, and hold the cap in place with nails, but do not set them (see drawing at right). If the post is steel, place it on the anchor bolts and plumb it. Using the top flange as a guide, drill pilot holes and start the lag bolts, but don't tighten them.

When everything is set, begin lowering the jacks 1/16 inch per day until the girder is settled firmly on top of the post. Resist any temptation to hurry this process. Finish securing the post by nailing through the post cap and metal post base or insert if you're installing a wood post, or by tightening the bolts on a steel post.

Continue to lower the jacks until the new post has assumed the full weight of the girder and the jacks can be removed.

How to reinforce a girder

If your sagging floor is caused by a wood girder that has rotted or is infested with termites, you'll need professional assistance to replace it.

Should the girder appear to be sound—no discoloring or evidence of rot or insect damage—it may be possible to level the floor without replacing the girder. A girder that sags between supports can be jacked up to level and an additional post added as a permanent support (see "Understanding the basic jacks," page 89). If the sag is over an existing post, the footing may have settled, and footing and post should be replaced.

Another alternative to replacing a sagging girder is to jack it up to level the floor and then reinforce it with steel plates or channels—their flat sides turned to the wood—on both sides of the beam. The steel should be bolted in place through the girder. An engineer should be consulted to size the steel and determine bolt spacing. This approach is also practical for strengthening a sound girder in order to add usable space by removing a post.

With proper counsel and help from friends, a talented handyperson can handle the reinforcing of a girder, but it is hard work. If it's necessary to relocate plumbing, ducts, or wiring, the job should certainly be turned over to a professional.

SQUEAKING FLOORS—ANNOYING BUT FIXABLE

Floors that talk back to you when you walk on them are trying to tell you something. The squeaks you hear are, technically, pieces of wood rubbing together, telling you that the subfloor may have separated from the joists or that loose boards need to be fastened down.

Eliminating squeaks is often relatively simple, once you know what's causing them. A little detective work will help you locate and identify the problem.

What causes squeaks

Squeaking floors can result from any one of a number of causes. Listed below are a few of the more common ones:

- Joists that are undersized or weakened by rot or termites, permitting excessive deflection (bending) and movement.
- Joists that have dried out and pulled away from the subfloor.

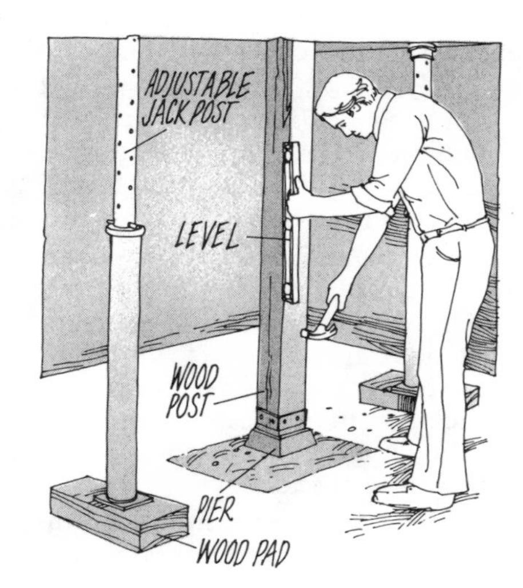

• Inadequate nailing of the subfloor to joists—common nails don't have the holding power of the annular ring nails or threaded nails recommended for securing subfloors.

• Settling of the house that has caused the subfloor to separate from joists.

• Weak or inadequate bridging between joists.

• Poorly manufactured wood strip flooring with undersized tongues that don't fit snugly into the grooves.

• Warped floor boards that rock when they're walked on.

• Sleepers that have worked loose from the concrete slab below them, permitting movement of the floor.

Locating the squeaks

Squeaks in floors with finished wood surfaces can originate in surface areas or in the subfloor. In floors with flooring materials other than wood laid directly over the subfloor, the squeaks will come from the subfloor.

In a typical frame home with exposed joists (visible from the basement or crawl space), it's easy to locate the area of the offensive squeaks. In homes where joists are not exposed or when it's the upper floors that are making noise, locating and correcting squeaks can be considerably more complicated.

If joists are exposed, watch from below while another person walks across the floor above; you should be able to detect the probable cause of the squeak—unusually springy floors, excessive deflection of the joists, or simple movement between joists and the subfloor.

Check the bridging between joists to see if it's firmly in place (see drawing on page 89 for the two most common types of bridging).

If you see or feel movement in the subfloor or joists, be sure to make all the routine checks for structural problems to see if the problem is traceable to rot, insect damage, or structural weaknesses (see "Locating the cause," page 87). Get help from a professional if you suspect serious structural problems.

Getting rid of the squeaks

Following are some relatively simple remedies for squeaking floors. Try whichever one seems easiest and most appropriate for the problem you've identified. If your first efforts are not successful, move on to what you believe to be the next logical step.

Shims and cleats eliminate movement. Simple wood shims can be used to eliminate squeaks caused by movement between joists and the subfloor. Locate those areas under the floor where movement can be detected. Tap shims lightly into the gaps between the joists and the subfloor (see drawing), but take care not to drive them in too forcefully or they'll further separate the subfloor from the joists.

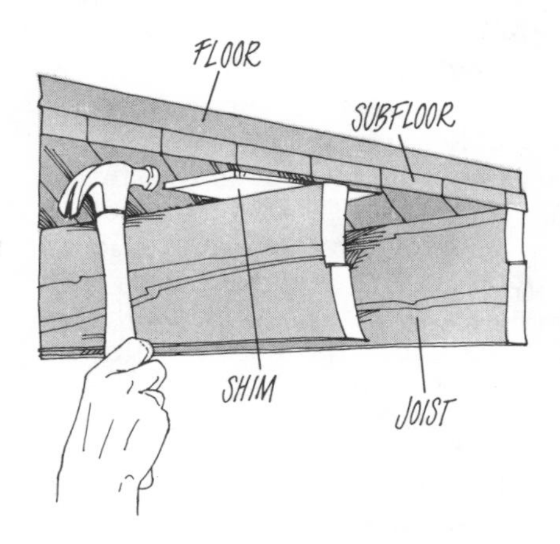

If the subfloor is made of individual boards laid diagonally, movement in an individual board or between boards can also be stopped with shims. If there are several loose boards, place a length of 1 by 4 or 1 by 6—it's called a cleat—against the joist and the loose boards; prop the cleat in place with a piece of 2 by 4. Tap on the 2 by 4 to hold the upper edge of the cleat snugly against the subfloor (see drawing). Then nail the cleat to the joist with 8-penny common nails. Make sure that no one is standing on the floor above while you install the cleat.

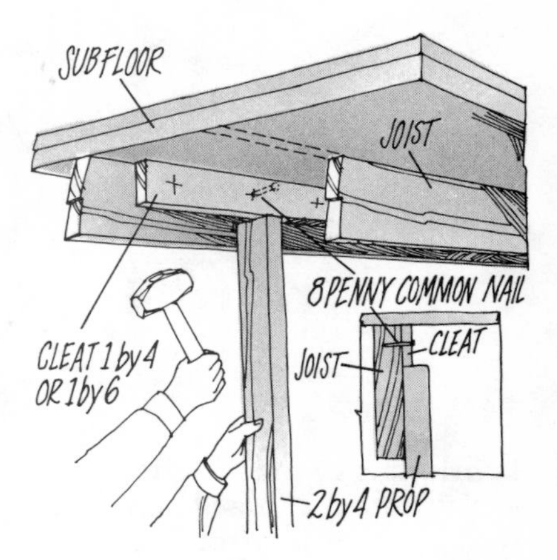

Plywood panels used for subfloors are commonly laid with their long sides at right angles to joists; joints between short sides of panels fall directly over joists. If you observe movement between panels in the space between joists, cut a 1 by 4 or 2 by 4 to fit under the joint between panels and attach it to the plywood with screws. This will eliminate movement and stabilize the panels.

Try tapping out squeaks. If you find that squeaking is limited to isolated areas of a wood strip floor, another simple approach is to try to eliminate the squeaks by tapping them out.

Take a piece of 2 by 4 about 1 foot in length. To avoid marring the floor, wrap the 2 by 4 in an old towel, or tack carpeting to one face of it. Then place it over the area that squeaks, at a right angle to the wood strips and, moving it in a rectangular pattern around the area, tap it sharply with a hammer (see illustration on next page). Avoid hammering over the same spot with too much force—this can damage tongue-and-groove flooring.

Lubricants reduce friction. Several different kinds of lubricants that reduce friction between boards can be applied to the surface of the floor to eliminate squeaks. This is a useful remedy for squeaking floors that are inaccessible from below. These are the possibilities:

Graphite. Powdered or liquid graphite squirted between boards will work its way down into the tongue-and-groove joints. Use it very sparingly: graphite is a messy substance if tracked across a floor.

Talcum powder. Dust the cracks between boards with talcum powder and wipe away any excess with a barely damp cloth or sponge.

Floor oil. Apply a liberal coating of good floor oil over the squeaking area and wipe up any excess with a dry cloth. Oil that soaks down into the cracks will expand the wood and tighten the flooring to eliminate squeaking.

Mineral oil. Another lubricant, mineral oil, can be used in minimum quantities to help eliminate friction. A few small drops in the cracks between boards will be sufficient. Too much mineral oil can stain the surface of a floor.

Drive glazier's points into cracks. Squeaks not vanquished by lubricants can often be silenced by glazier's points driven into the cracks between boards. Put graphite on the points and drive them into place between the boards, using a hammer; then use a piece of scrap metal or the edge of a putty knife to make sure the points are well below the surface. Use no more than one point every 6 inches along the squeaking area.

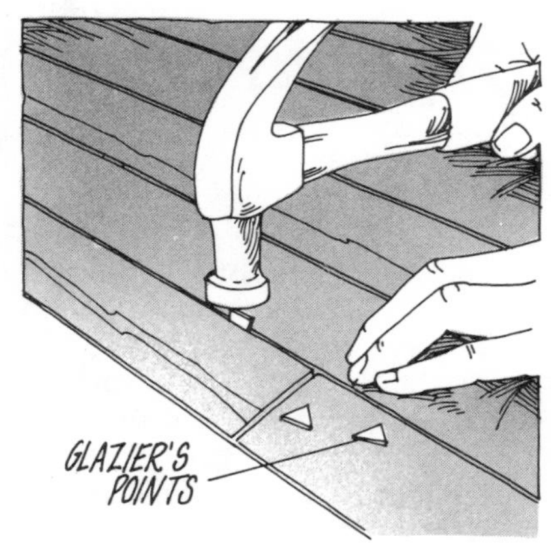

Pull boards tight from below. If individual boards are loose or bowed, the problem is more difficult to correct. If your floor is supported by exposed joists and it's possible to work from below, the best way to secure loose boards is to pull them tight with screws driven up through the subfloor. This makes it unnecessary to nail through the surface of the floor—an approach to avoid unless there are no alternatives.

Select woodscrews with round heads in a length that will not come to within ¼ inch of the finished surface when installed. Bore a hole up through the subfloor with a bit the size of the shank of the screw. Using another bit slightly smaller than the threads of the screw, drill a pilot hole directly into the underside of the flooring. Be careful not to drill too deep—no closer to the surface than ¼ inch.

Slip the screw through a large-diameter washer and up through the hole in the subfloor. As the screw is tightened, the loose board will be pulled snugly down against the subfloor.

If space permits, add glue. Most wood strip flooring has little or no space between boards. But if your strip flooring has wide enough cracks between boards—space enough to insert a putty knife—common white glue worked into the cracks will help bind the boards.

After adding glue, wipe away any excess. Then put some weight on the glued area and leave it in place overnight. Books, bricks, or any other heavy objects can be used, as long as you take care to protect the finish of the floor. Cover the glued area with a sheet of plastic before placing the weights—you don't want a prized book glued to the floor.

Surface nailing—a last resort. If working through a strip floor from above can't be avoided—because lubricants haven't worked or the floor is inaccessible from below—you can secure loose boards by nailing.

Using annular ring nails, drive the nails at a slight angle through the surface flooring and into the subfloor and, when possible, into a joist. If the flooring is hardwood, drill pilot holes slightly smaller than the diameter of the nails. This will reduce the risk of splitting boards and make it easier to countersink nail heads.

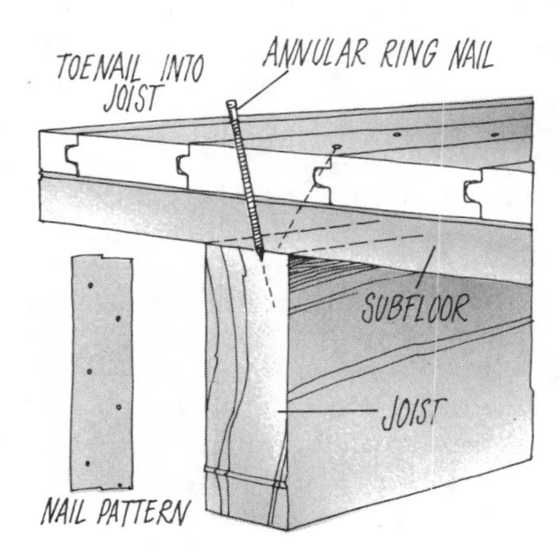

When the nails have been driven and countersunk, select a wood putty in a color that matches the finished floor, and fill the holes.

If a strip floor is covered by carpeting, finishing nails can be driven down through the carpeting. Take care, though, to avoid damaging carpet fibers with your hammer.

SQUEAKING STAIRS—SEVERAL SOLUTIONS

Probably more common than squeaks in floors, squeaks in stairs are caused by the same basic problem—wood rubbing against wood.

The solution is either to eliminate movement between treads and risers or to lubricate squeaking areas. Locating squeaks in stairs is, of course, easy. Simply move from step to step, pausing to rock back and forth on each step. Test the center of each tread first, then the ends.

If the undersides of the stairs are exposed—inside a closet, for example—eliminating squeaks

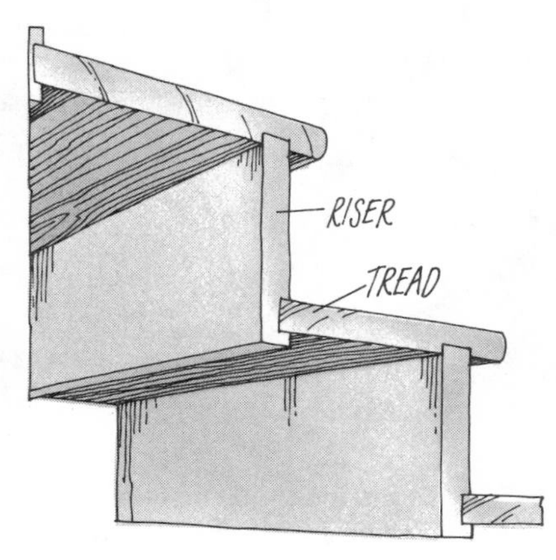

can be relatively easy. If your stairway is carpeted, being able to work from below can be particularly useful. When you don't have access from below the stairs, you'll have to work from above.

For cosmetic reasons, naturally, it is preferable to work on stairs from below; so among the following suggestions for silencing squeaky stairs, the remedies that can be applied from underneath are presented first.

Wedges make the simplest solutions to squeaking stairs. You just drive them between the riser and the tread, from below or above. While a helper puts weight on the squeaking step, look for any noticeable movement. Wherever there is play between the tread and the riser, you'll need to install wedges.

Wedges are best driven up from below, between a riser and the lip of the tread above it. The wedges should be small—an inch or two long; you can whittle them off a board or shingle, using a pocket knife. Dip the tip of each wedge into a common white wood glue. While the step is weighted, take a block of wood, place it against the blunt end of the wedge, and tap the block to drive the wedge snugly into place. But don't overdo it—you don't want to pry the stair and riser apart. Two or three wedges should be adequate. When the glue has dried, use a utility knife to cut off the protruding ends of the wedges, if you wish.

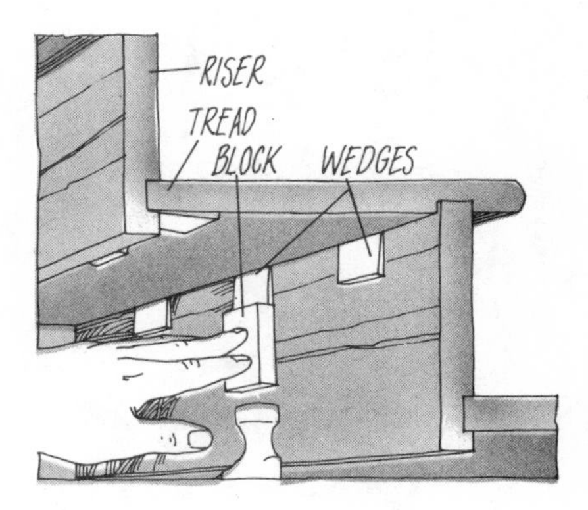

Where it's possible to work only from above the stairs, you can easily insert wedges between a tread and the riser above it. Here, too, coat the tip of each wedge with glue and carefully cut off the end when the glue has dried. When driving wedges, al-

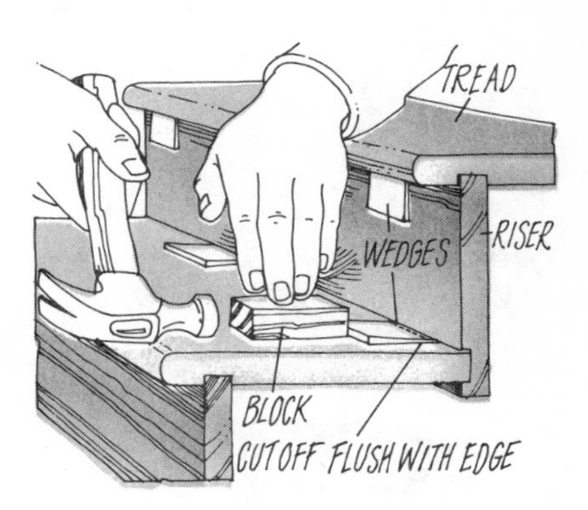

ways use a block of wood to provide a larger hammering surface and to keep from marring the stairs.

Wood blocks work better than wedges where you have access to the underside of the stairs. Glue 2 by 2 wood blocks tightly to the tread and riser where they join, and secure the blocks with woodscrews in both directions. Installing the screws will be easier if you predrill screw holes in the blocks. Be sure to select screws that are not too long—they should come no closer than ¼ inch to the surfaces of the treads and risers.

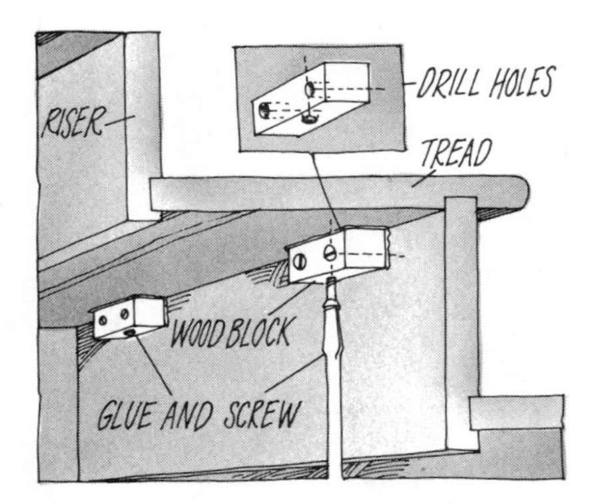

Shelf brackets or metal angles are easier to install than wood blocks—again, if you can work from below—though not as effective. The kind commonly available at hardware stores can

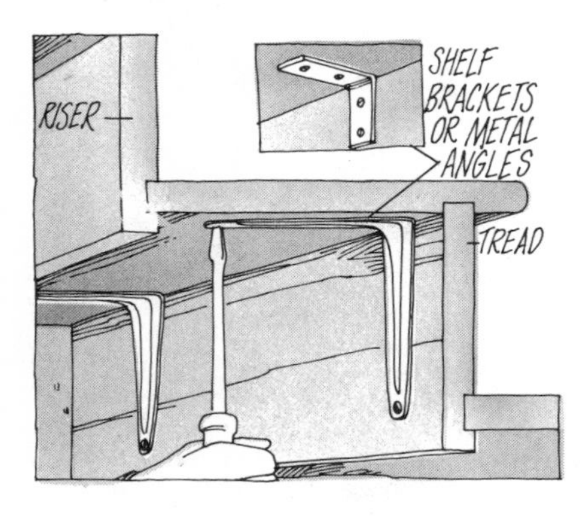

be used. Choose short woodscrews that will get a good bite but come no closer than ¼ inch to the surface of treads and risers.

When there's no alternative, molding makes a practical remedy that can be installed from above the stairs. Quarter-round or any decorative molding can be glued, then nailed to the angles between treads and risers to eliminate movement. The molding should be nailed to both surfaces.

Though you may need molding on only one or two stairs, you should install it on every step to

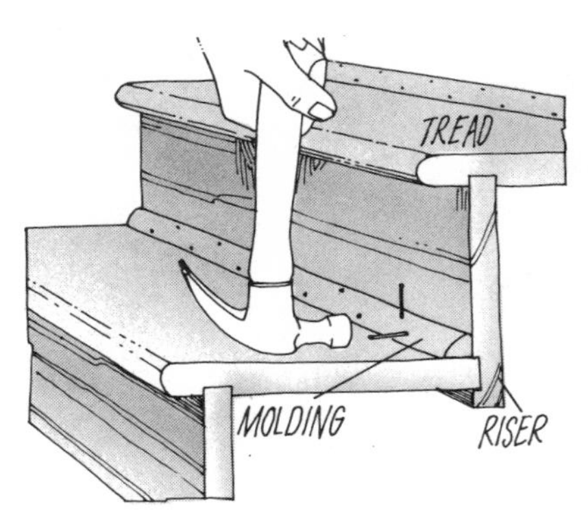

give the stairway a uniform, finished appearance. For the same reason, all nail holes should be filled, and the molding finished. Molding can be finished before it is cut and installed. This way, it only needs touching up once it has been nailed in place.

Lubricants used on squeaky floors—graphite, talcum powder, floor oil, or mineral oil—can be used sparingly in the joints between treads and risers to eliminate friction. This solution can be applied only from above, and may prove to be temporary. But since adding lubricant is a simple task, it can be repeated when necessary. (For details on using lubricants to eliminate squeaks, see page 94.)

Common white wood glue can be worked into the joints between treads and risers to bond them. Obviously, this work is best done from above and is effective only where the movement between pieces of wood is minor.

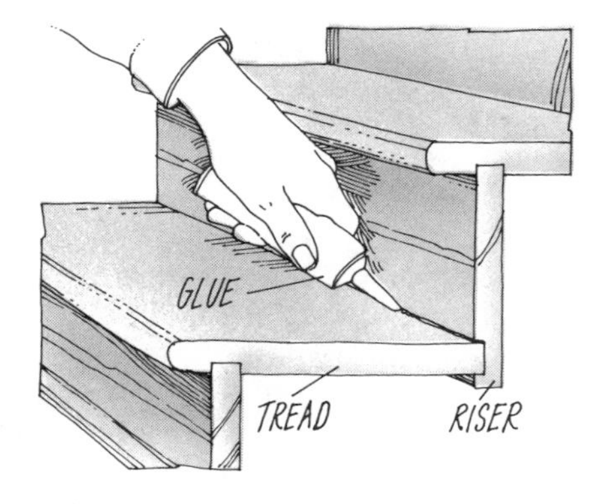

Work the glue into the crack with a putty knife or similar tool, and wipe away any excess glue immediately. Avoid walking on the step until the glue has dried.

REFINISHING WOOD FLOORS—EXACTING BUT REWARDING

Few improvements to a home are as satisfying as newly finished hardwood floors. Rich, natural wood tones in a floor enhance any room and bring out a home's best features. Synthetic floor finishes on the market today offer long life and easy maintenance, with the result that wood floors no longer need daily attention.

Wood floors installed in new homes are sanded to prepare them for sealing, staining, and finishing; floors newly installed in older homes get the same treatment. But the most common reason for refinishing is simply to rejuvenate a worn and damaged floor, or perhaps to give new life to one that's been hidden under carpeting for many years.

Refinishing a wood floor is a project that many homeowners consider doing themselves, if only to save the cost of having the job done by a professional.

The specific equipment needed—a drum sander, a disk sander, and a buffer—can be rented. Still, the work requires considerable patience and care; a single misstep can cause irreparable damage. And a refinishing project is likely to be time-consuming, messy, and disruptive to your general household comfort.

So before undertaking a complete sanding and refinishing job, investigate the wood reconditioning products available at hardware stores and home improvement centers. If you do decide that a thorough refinishing is in order, you have a third alternative to doing the complete job yourself or paying to have it done.

Sharing the work. Part of the cost of a professional refinishing job covers the preparatory work that must be completed before the actual sanding begins. But this is work that you can pretty easily do yourself.

In general, it will involve just taking up molding from around the baseboards, removing floor grates and other fixtures, and generally cleaning the surface of the floor. It might also include driving protruding nails below the surface with a nailset and filling the nail holes and any other gouges or dents with wood putty.

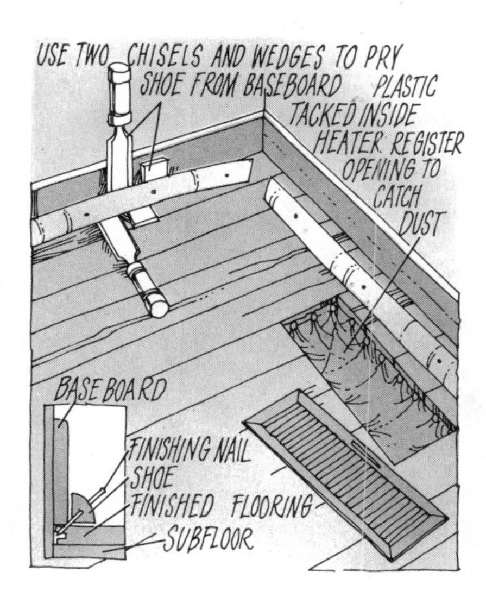

So as an alternative to having the complete job done for you or doing it all yourself, get a couple of professional estimates for sanding and finishing only. The estimates, or "quotes," should cover all materials and labor necessary to complete the job; you can get everything ready before the professional appears.

Inspect and repair floors before refinishing

Before you get on with the serious business of refinishing a floor, *all* floor problems should be corrected first—not only surface blemishes like damaged boards, but structural problems as well. The whole floor area, from below as well as from above, should be given a good examination (see "Locating the cause," page 87).

Sagging floors, squeaks, and loose subflooring are problems that are likely to get worse in the future. There's no better time to correct such problems than just before your floors are refinished. You'll find suggested remedies starting on page 88.

Choosing a finish

Floor finishing materials are of two general types—penetrating sealers and surface finishes.

Penetrating sealers actually penetrate the pores and become an integral part of the wood. The finish wears only as the wood wears. It can be retouched in heavy traffic areas without creating a patched appearance.

These sealers can be used clear or tinted, or purchased in premixed stain colors. Normally, two coats are required. Buffing a penetrating sealer while still wet will result in a satin sheen. A final coat of paste or liquid wax—but not a water-base wax—is recommended.

Some types of wood require a sealer before the finish coat; discuss this with your paint dealer. If you do plan to use a sealer as a first coat under a surface finish, check with your dealer to make sure the two products are compatible.

Surface finishes provide a clear coating over stained or sealed wood. In general, polyurethanes have replaced traditional finishes like varnish, shellac, and lacquer. Polyurethanes are blends of synthetic resins, plastic compounds, and other film-forming ingredients that produce an extremely durable, moisture-resistant surface. High gloss or satin polyurethane finishes are available.

Though some makers of urethanes claim that no waxing is necessary, hardwood flooring manufacturers believe that waxing over a polyurethane finish will produce a better appearance and give longer wear.

If you're going to refinish a softwood floor, it may be necessary to seal it before staining, or to use a sealer mixed with stain, to achieve a uniform color. Ask your dealer what products to use and how to use them.

If you're refinishing a wood block floor made of prefinished tiles, ask the supplier or manufacturer to recommend compatible sealers and finishes.

Many tiles are made with rounded or beveled edges—making it impossible to remove the old finish between tiles by machine sanding. If you're refinishing such tiles, you'll want to choose a finish that will match or blend with the old finish, which will remain on the tile edges.

Tools and supplies

The basic heavy-duty equipment you'll need to refinish a floor is available at most equipment rental companies. Often used by people who aren't wholly familiar with its operation, this machinery is sometimes abused. Take care to check over all equipment you rent, and ask that it be tested before hauling it home.

Floor sander. The workhorse you'll need is the basic drum sander, designed specifically for stripping old finishes off wood floors and producing a finely sanded surface ready for sealing and finishing.

Drum sanders come in several commercial models, but look

for one that has a tilt-up lever that makes it possible to raise the drum off the floor without having to lift the machine. Be sure to get a machine with a dust bag. This won't eliminate dust, but it will cut down the quantity. Check to see that the machine operates on 110 volts and doesn't require 220-volt outlets. Most sanders have three-prong plugs, so if you don't have outlets that take three-prong plugs, you'll need an adapter. If you have to use an adapter, make sure it's grounded properly.

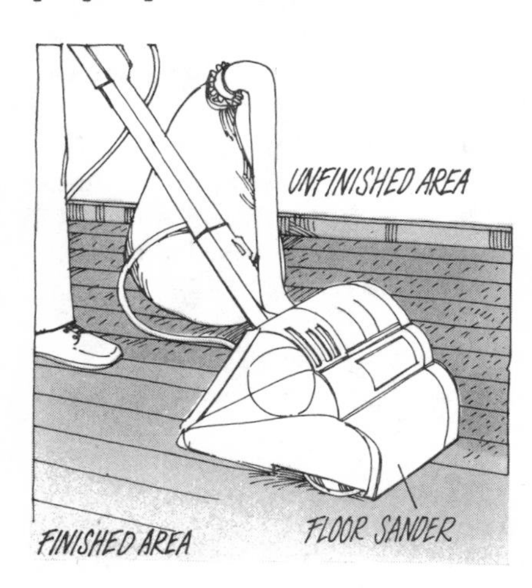

Before you take the sander home, ask the rental agent to show you how to load the drum and make sure he provides the wrenches needed; it's a relatively simple procedure if you have the right wrenches and know which wrench to turn in what direction. Also ask for tips on operating the machine. Despite its size, a drum sander is a delicate piece of equipment to operate.

Edging sander and hand scraper. An edging machine is a type of disk sander necessary for reaching areas (next to walls, for example) that can't be reached with a drum sander. You'll also need a hand scraper for cleaning out corners and reaching into other tight areas—such as around radiators.

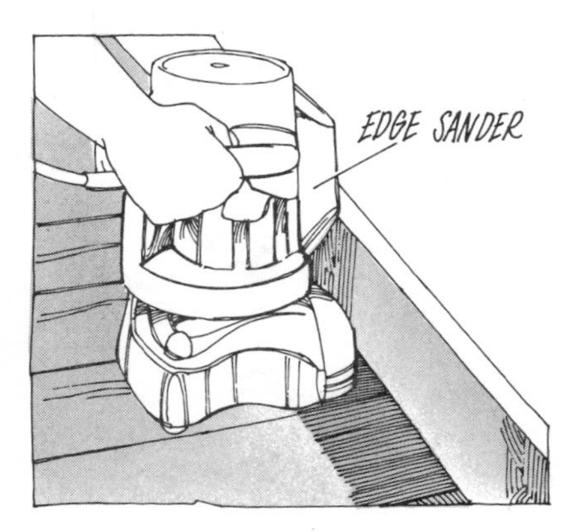

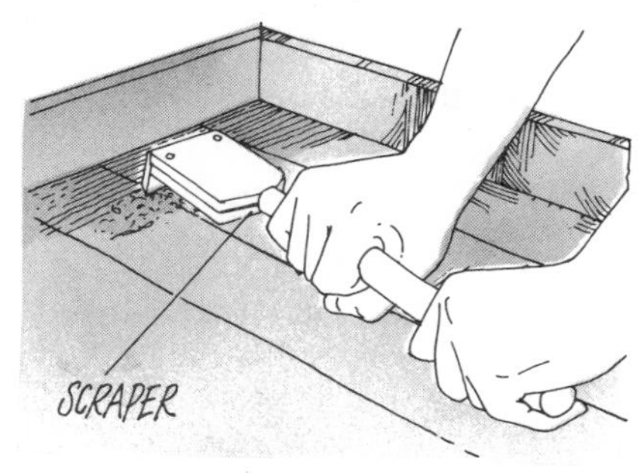

Buffer. A professional floor buffer, the type with a single large revolving pad, is needed to buff the floor with steel wool after the stain and/or sealer have been applied.

Sandpaper. Because three sandings—called "cuts"—are required to remove an old finish, you'll need three grades of sandpaper. Depending on the finish you're removing, the first sanding should be done with 40-grit to 20-grit paper (the lower the number, the coarser the paper).

A 20-grit paper should be used to remove old paint, or where a floor is particularly rough. A 30-grit paper is about right for removing old shellac or varnish. A 40-grit paper is good for sanding a herringbone or wood block (parquet) floor, or a wood strip floor in good condition.

A 60-grit paper is needed for the second sanding, and 80 or 100-grit paper for the final sanding.

To sand the floor of an average room, about 14 by 16 feet, you'll need about two sheets of each grade of paper and four edger disks of each grade. Worn paper won't work as a substitute for a finer grade.

Other helpful supplies. When sanding a floor, you may need a hammer and nailset to drive down protruding nailheads, and wood putty to fill holes, dents, and gouges. Choose putty that will blend with the finish to be used or will take a stain. Professionals use a product called "goop-on" for filling.

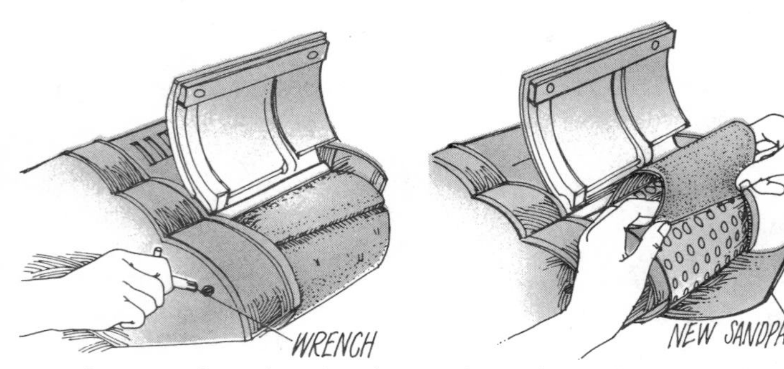

Use wrench to open slot in drum (sander must be unplugged); wrap sandpaper around drum, tucking ends into slot so paper is tight. To close slot and secure sandpaper, turn wrench in opposite direction.

Sanding a floor is a very dusty and noisy job. Anyone operating a power sander should wear a mask. Earplugs or earmuffs will help prevent a throbbing head and possible ear damage.

The dust kicked up by sanding manages to escape through unbelievably fine openings, so you'll want to seal doors to other rooms with masking tape or sheet plastic. Taking such care before sanding will save you a good deal of housecleaning later.

You'll also want to wear soft-sole, nonmarring shoes—and make sure that no one walks on your newly sanded floor. With the pores of the wood exposed, an unprotected floor will soil easily.

Sanding a strip or plank floor

As you prepare to get down to the work of sanding your floor, keep in mind that the first coat of the new finish should be applied the same day the sanding is completed. This will prevent moisture in the air from raising the grain in the raw wood, and dirt from marring the exposed floor. So plan your time accordingly.

To prepare the floor for sanding, take up the molding along the baseboards. If you remove the pieces of molding in sequence and number each piece with chalk or a pencil as you go, it will be easier to replace them when the floor has been refinished. Also remove all floor grates, door stops, and any other fixtures. Set any protruding nails below the surface of the floor, and fill the nail holes—and any other dents or gouges that you see—with wood putty. Finally, sweep the floor clear of dirt and debris.

In sanding strip or plank flooring, you'll always go with the grain—unless you're faced with an exceptionally rough floor or a floor with cupped boards. In such cases, the first cut (sanding) should be made on the diagonal; otherwise, the procedure is the same.

A floor sander going at full tilt can be an intimidating piece of equipment; handling it is not an easy job. But if you follow a few basic rules, you'll quickly gain confidence and get the upper hand. Keep in mind that *the objective is to take off the minimum amount of wood possible while removing all of the old finish.*

Making sure that the drum sander is unplugged, begin by loading the drum with the coarse-grade paper you've selected (see illustrations on facing page). When the paper is in place, plug in the sander and prepare for the moment of truth.

The first sanding. When you're ready to make your first pass across the floor, lower the drum slowly and move out, literally, the instant it touches the floor. *Do not allow the drum to bite into the floor before beginning to move.* If you do, you'll gouge the floor, leaving a permanent reminder of your error.

The forward rotation of the drum will pull the machine forward. Keep it in hand and move forward at a steady pace. As you approach the end of your first run, *be ready to lift the drum off the floor while the sander is still moving forward.* If you try to stop the sander, then lift the drum, you won't be able to act quickly enough to avoid taking a hunk out of the floor.

When you've completed one pass, shift the sander over to begin the next pass, with the drum overlapping the last cut by 2 or 3 inches. On the return trip, you'll have to pull the sander backwards; go at the same steady rate.

When you've finished sanding as much of the floor as you can with the drum sander, it's time to load the disk sander with the same coarse-grade sandpaper and sand the areas the drum sander was unable to reach. Work on these areas until the disk has removed about the same amount of material as was removed by the drum sander.

When you've completed the first sanding, set any protruding nails and fill all nail holes, dents, and gouges.

The second sanding. Load the drum sander with a medium-grade sandpaper and make a second cut over the whole floor. Then, with medium-grade paper on the disk sander, go over the edges once again.

If after the second sanding you discover exposed nail heads or uncover holes that need filling, set the nail heads and fill all holes with wood putty.

Between the second and final sandings, clean out any spots you were unable to reach with either of the power sanders during the first two sandings. Corners, areas around pipes and radiators, and door jambs have to be scraped by hand. When using a scraper, pull toward you while exerting downward pressure. If possible, scrape with the grain. Keep a file handy—the scraper should be sharpened frequently to make it an effective tool.

When you've scraped away the old finish, wrap a medium-grade sandpaper around a block of wood and go over the scraped areas. Finish off with a fine sandpaper of the same grade you plan to use for the final floor sanding.

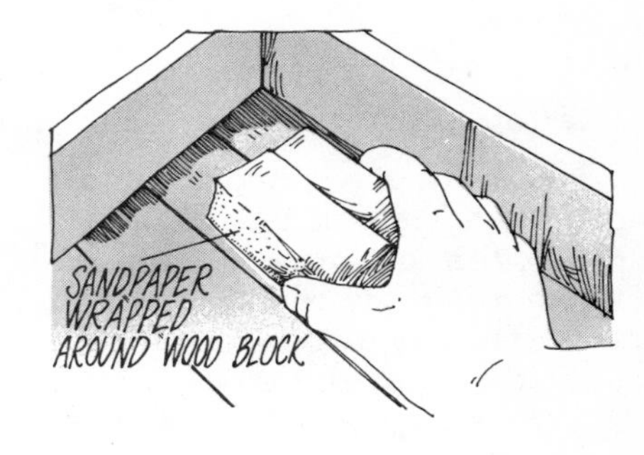

The final sanding. Using fine-grit sandpaper, go over the entire floor one last time with the drum and disk sanders.

Sanding wood block floors

Wood block flooring—whether of tiles or inlaid wood mosaic—is commonly called parquet, and it comes in a variety of styles and types of wood. The basic directions for sanding a strip or plank floor (see page 99) apply also to a wood block floor. The only difference is that *with wood block flooring, all three sandings should be done in a diagonal pattern*. This compensates for the fact that the wood grains in these floors run in several different directions.

Make the first pass with the drum sander from one corner of the room diagonally across the room at a 45° angle. The second sanding—the one with medium-grade sandpaper—runs across the other diagonal of the room to produce a crisscross pattern. The final fine sanding follows the direction of the first sanding.

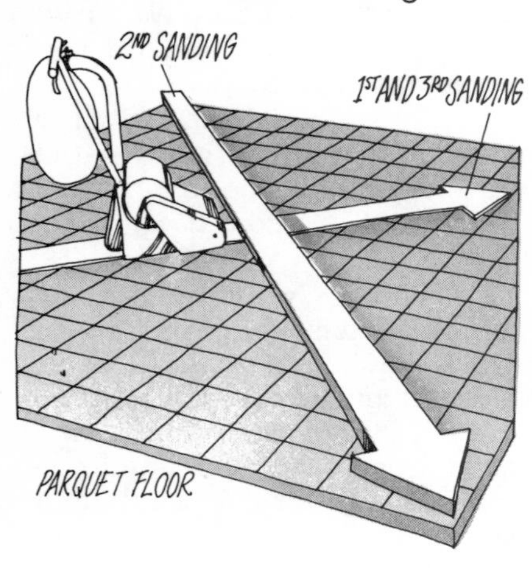

Cleanup is crucial

When the final sanding has been completed, clean up the wood dust with a vacuum cleaner. Make a final sweep of the floor with a barely moist mop or, better yet, a tack cloth.

Check the surface carefully for any blemishes that may have escaped notice earlier, and take care of them now. If any nail holes are showing, fill them with wood putty selected to match the final floor finish color, and sand lightly with a fine grade of sandpaper.

In disposing of the debris from the dust bags of the power sanders, do not dump the dust directly into a closed garbage can. The dust will be warm and will contain wax and varnish residue—it could catch fire. Allow it to cool before disposing of it in closed containers.

Applying the finish

When you have a perfectly clean and exposed floor under your clean feet, you can go ahead and begin the process of applying the final floor finish.

Remember: the first coat should be applied the same day the final sanding is finished. This will keep the exposed wood from getting dirty and absorbing moisture that may raise the grain.

There's no general agreement as to which finishing products are best. You can decide whether or not to use a stain, and exactly what type of finishing coat to apply, according to the specific results you want.

Regardless of what products you've selected, read through the manufacturer's directions carefully before applying any finish. It's wise to test the products you've chosen in a closet or out-of-the-way place before applying them to the entire floor.

We found no single preferred procedure for finishing a floor. Among professionals, the method will vary according to the products being used and the person doing the job. What follows is a step-by-step description of one typical procedure. It involves using a stain and then applying either a penetrating sealer or a hard surface finish (the characteristics of the two are different—see page 97).

Applying a stain. Pour all the stain you'll be using into a bucket and mix it thoroughly. Dip a clean, dry rag (of lint-free fabric) into the stain and spread it liberally over the floor. Near the walls, apply the stain with a clean brush.

Wait 5 or 10 minutes to allow the stain to penetrate the pores of the wood; then use clean rags to wipe up any excess stain. Let the floor dry overnight.

When the stain has dried, buff the floor with #2 steel wool. Follow this with a particularly thorough vacuuming of the floor. Now the floor's ready for a coat of either penetrating sealer or hard finish.

Applying a penetrating sealer. In general, a penetrating sealer is a free-flowing liquid that can be applied with a clean rag, brush, or roller. The manufacturers of specific products may have specific recommendations.

Apply sealer liberally, allowing it to flow into the pores of the wood. Start in a corner or next to a wall to avoid having to walk over wet sealer. After the sealer has had ample time to penetrate (check the manufacturer's instructions), wipe up any excess with dry rags and let the sealer dry for the length of time recommended by the manufacturer. Drying time may be affected by humidity and temperature.

Decide how many coats of sealer you should apply to a newly refinished wood floor according to the sealer manufacturer's suggestions and the results you want to achieve.

Applying surface finishes. Typical surface finishes—polyurethane, varnish, lacquer, and shellac—are applied like paint. They produce a surface with a different texture from that left by penetrating sealers.

Polyurethane has become the dominant floor finish used today because it provides a hard, plasticlike finish that is much easier to take care of than other surface finishes.

To apply polyurethane, use a clean brush to put down a coat of finish along the walls and around obstacles. Then use a long-handled paint roller with a mohair roller to apply the finish evenly over the rest of the floor. Work with the grain when possible.

Typically, two coats are required. Allow the first coat to dry (follow the manufacturer's recommendations); then use a floor buffer equipped with #2 steel wool to smooth the surface. Corners and hard-to-reach areas should be smoothed by hand.

Clean the floor thoroughly with a vacuum cleaner. Then go over it with a barely damp mop or rag to pick up fragments of steel wool and dust. Be thorough—any dirt left on the surface will be sealed in when you apply the second coat of polyurethane.

When the floor is clean and dry, apply a second coat of finish, working across the grain when possible.

No longer widely used, varnish, shellac, or lacquer are applied in much the same manner as polyurethane. Check the directions of the manufacturer before applying any of these special finishes.

The "finishing" touches. Allow the floor sufficient time to dry—plan on about 72 hours—then replace the molding, grates, and other fixtures that were removed.

If you've used a penetrating sealer, the floor will require waxing. Select a wax made for floors, and follow the manufacturer's recommendations for applying it. Polyurethane is sold as a finish that needs no wax, but many homeowners prefer the look obtained by applying a coat of wax.

For additional information on caring for your newly finished floor, see the special feature "Caring for a Wood Floor," page 56.

FLOOR SURFACES—REPAIRS THAT WORK WONDERS

Many floor problems are only skin deep and can be taken care of with minor repairs. For example, replacing a square of resilient tile or patching a sheet vinyl floor can be relatively simple. It's a little more complicated to replace sections of damaged hardwood flooring or individual ceramic tiles, but the job can be tackled by the homeowner with average building skills.

In this section you'll find out how to make surface repairs in wood strip, plank, and block floors; resilient flooring; ceramic and masonry flooring; and carpeting.

In most cases, the cause of the damage will be obvious—a burn, a tear made by moving a heavy appliance, or a crack produced by dropping a heavy object. If the cause is not obvious, take the time to examine the entire floor surface, and check its supporting structure from below to see if the surface damage has been caused by structural problems. Cracked ceramic tile in a frame home can be caused by movement in a floor that is no longer structurally sound. Deteriorating wood may be the result of a plumbing leak or excess moisture soaking into a floor.

Detailed instructions on how to check for structural problems begin on page 87; read through them carefully and make a thorough examination. A description of the steps necessary to correct various structural problems begins on page 88.

Replacing tongue-and-groove flooring

Replacing individual boards in a wood floor is a project that should be undertaken only if some simpler remedy—such as sanding and refinishing—won't work. It doesn't require exceptional skill to do the work, but it does take patience and finesse. It may be very difficult to match new pieces of wood and new finishes with the surrounding floor.

Ideally, this type of repair work will be one step in an overall floor refinishing project in which the entire floor will be given a new, uniform finish—this will solve the problem of matching. (See "Refinishing Wood Floors," page 96.)

Tools you'll need. For either repair method described below, you'll need a few basic hand tools—a combination square; a sturdy, sharp 1-inch chisel; a hammer; and a pry bar with a curved end.

For cutting out a rectangle, you'll also need a portable power saw with a blade that can be adjusted for depth of cut. A portable electric drill with a ½-inch bit will be useful for taking up individual boards if you're removing them in a staggered pattern.

Matching wood can be problem. It may be impossible to buy replacement flooring that matches your floor perfectly. Plan to take a sample to your lumber dealer to find a suitable replacement. If your floor is made of prefinished flooring, a match may be easier to find.

For a detailed look at the type of flooring you may be dealing with, refer to page 5 in the introductory section of this book.

Two ways to replace damaged boards. There are two common approaches to replacing boards in random-length hardwood floors. The first is to cut out a rectangle, remove the damaged boards, then replace them with boards of equal length (see the step-by-step illustrations on the next page). This approach is satisfactory for areas that will be covered by a rug or furniture. The second and slightly more difficult method is to remove boards

(Continued on page 103)

Replacing boards in a rectangular pattern

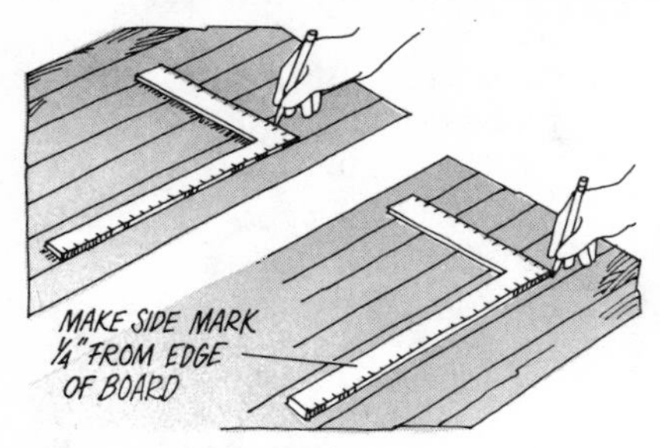

Mark work area with square and pencil. Use lengthwise crack to align square for end mark; make side marks ¼ inch away from cracks so saw won't hit nails.

Make end cuts by positioning toe of sole at midpoint of end mark. Turn on saw; before moving ahead, slowly lower blade until sole rests flat on floor. Work from center out to each edge.

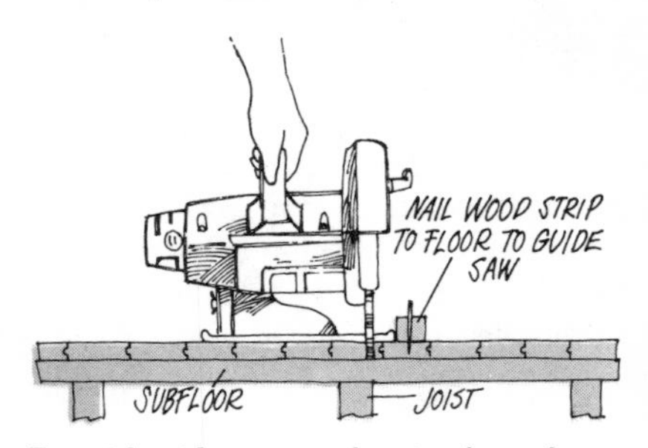

To guide side cuts, tack strip of wood parallel to pencil line and away from area to be removed. Work from center toward ends. Adjust saw blade if shoe rides on strip.

Using hammer and chisel, complete cut to subfloor. Keep beveled face of chisel toward boards to be removed. Take care to make clean cuts at corners, so boards don't splinter into good area.

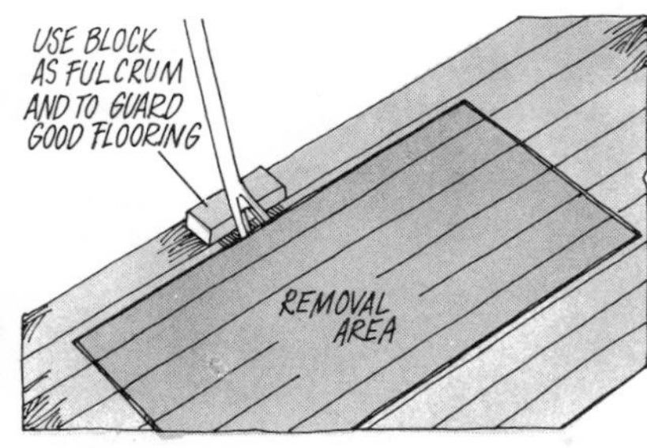

Lift boards with pry bar, starting at midpoint of side cut. Use small wood block for leverage and to avoid marring any part of surrounding flooring.

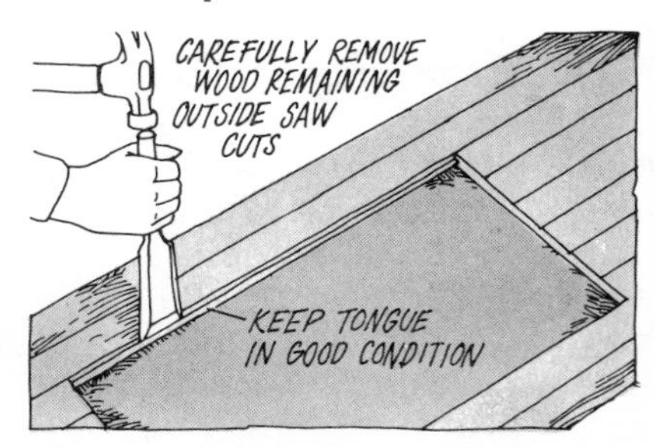

Cut away ¼-inch edges left outside saw cuts using hammer and chisel. Work slowly to avoid damaging adjacent boards. Set exposed nail heads.

Measure for exact length of each new board with steel tape marked in 1/32s or 1/100s of an inch. Start from 2-inch mark if tape end doesn't give clear mark to start first inch. Hold tape taut.

Score pencil marks on new boards with light saw cut at 90° angle. Make cuts on waste side of each pencil mark so saw kerf doesn't cause new board to be short.

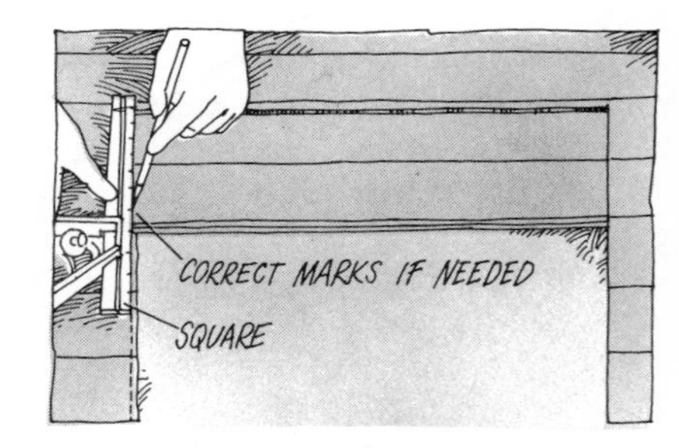

Lay board, scored face up, over work area. Fit one end in tightly; then check other end and mark for tight fit. Make any adjustment required; saw board.

Blind nail new boards as shown after sliding grooves of new boards over tongues of boards already in place; work steadily toward opposite edge.

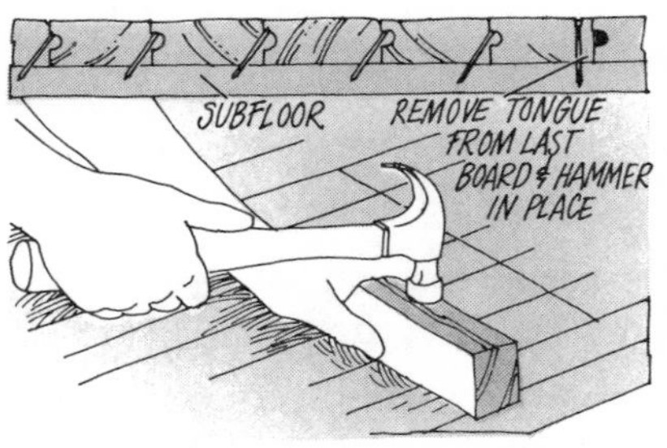

Remove tongue from last new board. Sand cut edge, if necessary, so board fits tightly; set in place. Using hammer and wood block, tap board down. Face nail board to secure it.

Set nails; fill nail holes and any obvious mismatches of end joints with wood putty. Then sand, stain, seal, and finish new boards to match finish of surrounding floor surface.

...Continued from page 101

in a staggered pattern (see the step-by-step illustrations below). This produces a less noticeable repair and is best for an open floor area.

Replacing wood blocks

A damaged square of wood block flooring is not difficult to replace; the toughest part of the job might well be finding the right substitute.

Matching new to old. You may have a problem matching the finish of a replacement block with the old floor—especially if the floor was sanded and finished after installation. Reserve a piece of the damaged wood block you remove to show to a flooring materials dealer; this will help you find a match.

Matching prefinished block flooring is somewhat easier, particularly if a few extra pieces were set aside when the original floor was installed. Even if you don't have any extras, you may be able to find a suitable match through a flooring dealer.

Because wood block flooring is often cemented in place with a mastic that can be softened with solvent, you may be able to remove an original square from a closet or some other unseen area and swap it with a new block that's not a good match. But removing a piece of block flooring without damaging it requires patience and care, and this procedure should probably be attempted only as a last resort.

Keep in mind, too, that wood block flooring is manufactured in many different forms and attached with any of several types of adhesives; before attempting a swap, ask your flooring dealer for suggestions on how to proceed.

Replacing boards in a staggered pattern

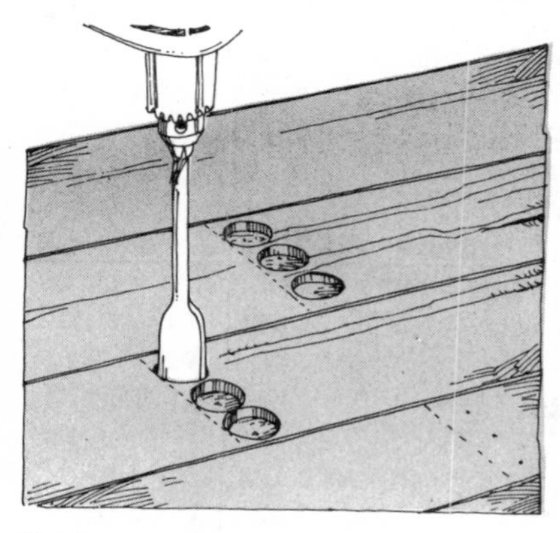

Use large spade bit to drill series of holes (do not drill into subfloor) at both ends of marked lengths of boards to be replaced.

Split defective area of each board using large wood chisel and hammer. Pound lightly to avoid splitting or cracking surfaces of adjacent boards.

Pry out split lengths of board until all are removed. Use small wood block for leverage, if necessary, and to avoid marring any part of surrounding flooring.

Slip groove of carefully measured and cut new board over tongue of board in existing flooring; use scrap of flooring and mallet or hammer to tap new board in place.

Blind nail new pieces tightly in place. If adjacent existing boards have separated, use thin shims to align edges of new boards with edges of old.

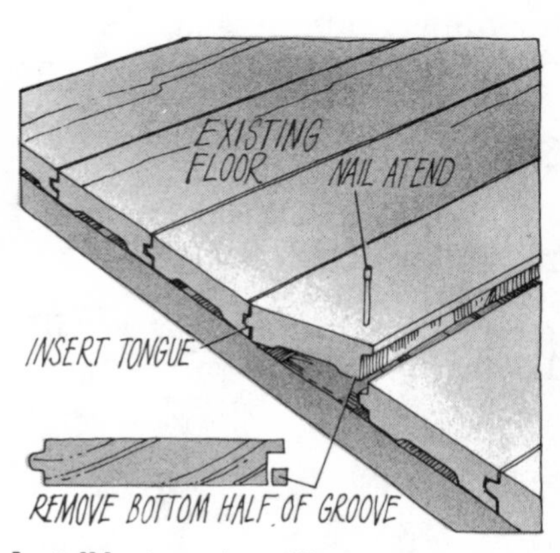

Install last new board by removing bottom half of groove, slipping tongue into groove of existing board, and pressing board into place; face nail each end.

Tools and supplies you'll need to replace wood block flooring are the same as are used to replace wood strips (see preceding section). In addition, you will need adhesive and solvent—ask your supplier for recommendations for your particular situation.

To remove wood block flooring, set the blade depth of your power circular saw to the thickness of the wood block. Make cuts near the edge of the damaged block, taking care not to cut into any adjoining squares (see illustration). Then take a hammer and

chisel and remove the damaged wood. Chip out as much of the remaining adhesive as possible—enough so that the replacement block will sit flush with the surrounding floor.

Replacing wood block is a simple matter. If you're installing a tongue-and-groove block, remove the bottom grooves (see illustration). Spread adhesive on

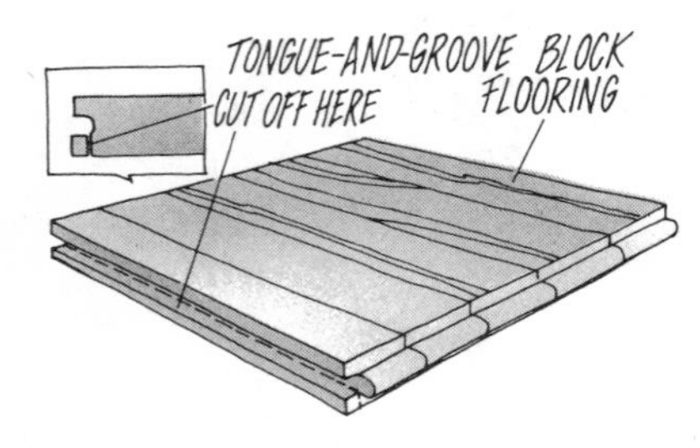

the bottom of the new block and set it in place. Tap it lightly, taking care not to mar the surface. If you get adhesive on any adjoining squares, quickly remove it with a solvent recommended by the adhesive supplier.

Replacing ceramic tiles set in thin-set adhesive

The qualities that make ceramic tile a desirable flooring material—its hard surface and rigidity—also make it susceptible to cracking. A heavy object dropped on tile can easily crack or chip it, and structural problems, like cracks in a concrete slab or settling in a frame house, can cause a series of cracks in a tile floor.

Another possibility is that structural shifting may leave individual tiles intact while opening gaps between tiles or causing grout to deteriorate. Sometimes, too, the grout around perfectly good tiles may need replacing when it becomes stained or cracked.

Simply chip out the old grout from the joints using a cold chisel and hammer; then scrub the joint surfaces with scouring powder or tile cleaner. Rinse the surface well. Apply the grout, following the instructions on page 71 or the directions on the package. Most tile suppliers sell grout in small quantities for replacement purposes.

Matching old tile. Though replacing a single ceramic tile or a few tiles is not difficult, finding a good match between replacement tiles and an old floor can be a problem. Colors may vary from one firing to another, even when the manufacturer uses identical glazes. That's why it's always a good idea to buy more than enough tile when installing a new floor, so you'll have extras to use as replacements.

If you aren't fortunate enough to have matching tile left over from the original installation, you may be able to find an acceptable substitute by checking with several flooring or tile specialty stores. Should you find nothing that does the job—a likelihood particularly if you're replacing a pattern that's been discontinued—it's time for a little creativity. Consider using complementary colors as replacements.

Tools and supplies. To remove damaged ceramic tiles, you'll need an inexpensive glass cutter, a common lever-type can opener, a hammer, a nailset or center punch, a combination square, a putty knife, and a small cold chisel or a "chipping" hammer (usually rentable at stores that specialize in ceramic tiles). A portable electric drill with a ¼-inch masonry bit will also come in handy. And wear goggles to protect your eyes while chipping out old tile.

If you're planning on replacing tiles of odd shapes or around pipes, bathroom fixtures, or similar obstructions, you'll need a pair of ordinary pliers or tile nippers, as well as emery cloth or a carborundum stone.

A few tools used to install new tile flooring may also come in handy (see "Tools and supplies," page 67). The extent of your need for specialty tools for installing replacement tiles will depend on the size of the area to be retiled.

Finally, you'll need adhesive and grout to cement the new tile in place. Your flooring materials supplier will be able to help you select the correct adhesive for the job and provide matching grout.

Replacing the tile is a fairly simple matter of loosening and chipping out the old tile, replacing it, and regrouting around the new tile. Step-by-step directions for removing damaged tile, installing replacements, and applying new grout are illustrated on the facing page.

Replacing damaged ceramic tile

Remove grout from joints around damaged tile using a lever-style can opener. Skip this step if grout joints are more than ⅛ inch wide.

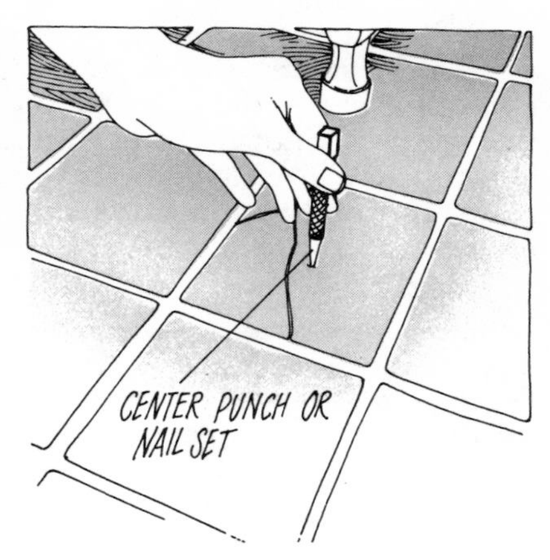

Punch hole through center of damaged tile using hammer and center punch or nailset. Pound gently so subfloor below isn't damaged.

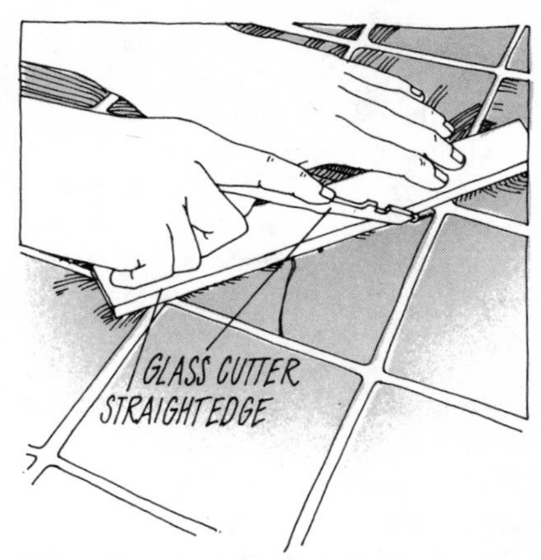

Use glass cutter and straightedge (hold it firmly) to score a deep X across face of damaged tile through center of hole.

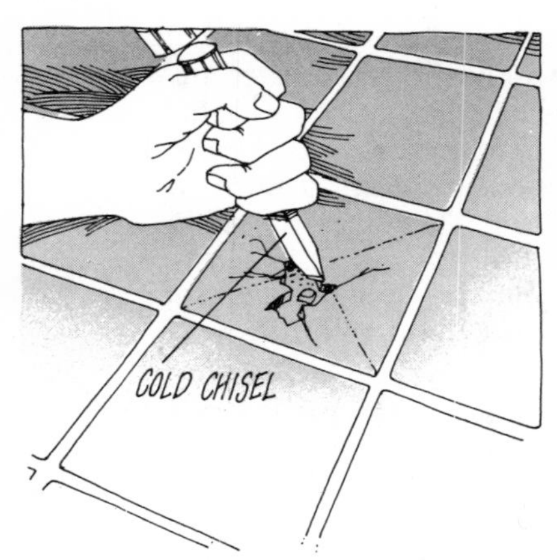

Starting at center, chip out old tile (and any remaining grout) using hammer and cold chisel; use light, rapid blows and work toward edges.

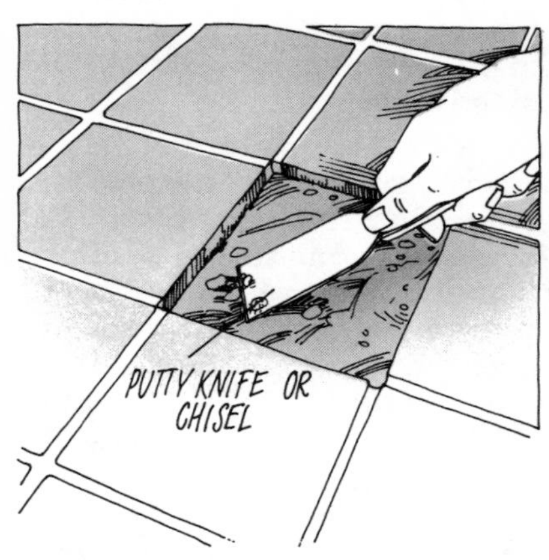

Clean subfloor, removing old adhesive and any bits of remaining grout. Use sandpaper to smooth rough spots and edges of surrounding tiles.

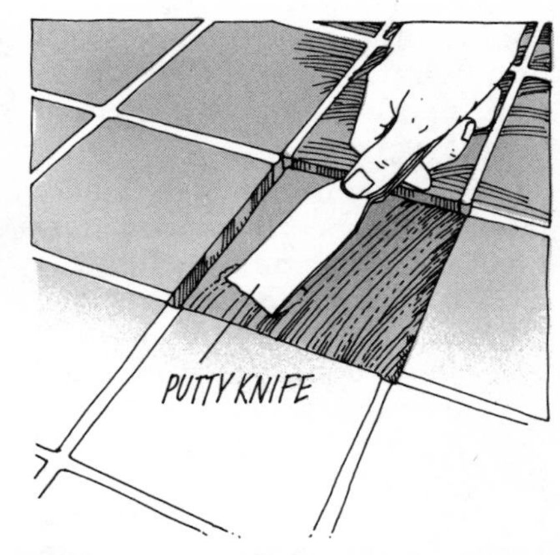

Build up surface, if necessary, with patching plaster so replacement tile will be level with surrounding tiles. After plaster dries, paint with latex primer.

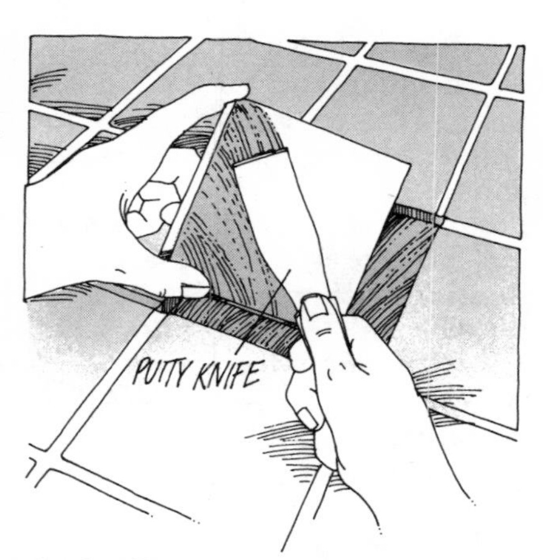

Apply adhesive to back of replacement tile, using putty knife. Spread adhesive smoothly following adhesive manufacturer's directions.

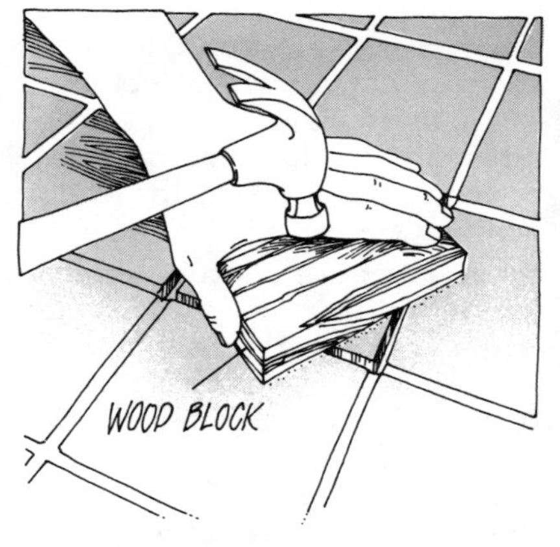

Tap tile gently in place using wood block and hammer. Allow adhesive to set for at least 24 hours before grouting. Detour traffic around new tile.

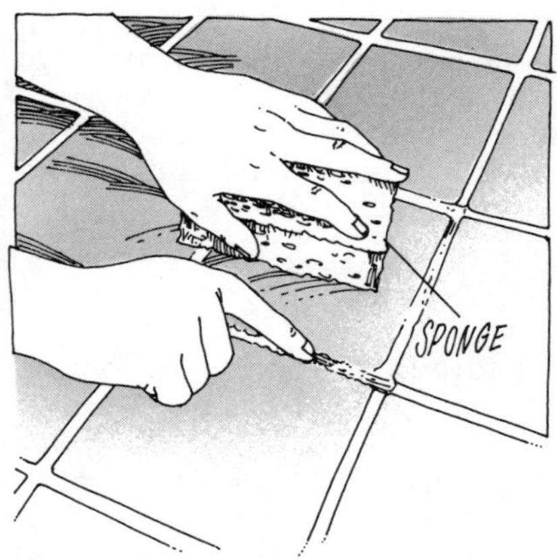

Apply grout to joints using damp cloth, sponge, or squeegee. Use your finger to smooth grout joints; clean off excess with damp sponge.

Replacing mortar-set tiles and other masonry flooring

Many ceramic tiles, as well as slate, flagstone, brick pavers, and other kinds of irregular or oversize masonry-type flooring are set in a thick bed of cement mortar. Their uneven surfaces make other adhesives impractical. Fortunately, these heavier flooring materials are so rugged, and mortar provides so strong a base, that they rarely need replacing.

It's impossible to remove such materials without breaking individual tiles or pieces. The only way to dislodge them is to put on a pair of safety goggles, take a hammer and cold chisel, and chip away. A power drill with a masonry bit may come in handy, but in any case the job requires patience and a good deal of elbow grease.

The area from which the old flooring has been removed must be cleaned out and enough old mortar cleared away to make room for a new bed of mortar. After the area has been cleaned out, the surface must be coated with a bonding agent (ask a flooring materials supplier for the proper product) before new materials are set in cement mortar.

For small projects, commercial mortar mixes are handy for installing replacement flooring; simply follow the directions of the manufacturer. For larger projects, see pages 72–76 for details on installing masonry-type floors. You may also want to refer to the *Sunset* book *Basic Masonry Illustrated*.

Repairing resilient flooring

Usually, all flooring materials classified as resilient—vinyl, vinyl-asbestos, asphalt, rubber, cork, and other similar compositions—are easy to repair. Following are step-by-step instructions on how to make the most common surface repairs. But it's a good idea, before going ahead with any repair project, to take the time to determine if the trouble on top of your floor is caused by more serious problems below the floor. So read through the following material completely and do a little sleuthing before undertaking repairs.

Look for the cause. If the cause of damage to resilient flooring isn't readily apparent, you should give the structure below a thorough inspection for structural weakness or damage from insects or rot. Some relatively minor surface blemishes can often be traced to more serious problems in the subfloor or supporting structure—problems that not only affect appearance but also threaten to cut short the life of the flooring.

A regular pattern of indentations, running for several feet or forming "T"s, may be caused by separations in the subfloor. Individual boards in a subfloor can shrink and leave gaps between boards; plywood panels may separate as the structure settles.

Small bumps that appear in the surface of the floor may be caused by nails that have worked loose. Over a period of time, movement in the structure can cause the subfloor to separate from the joists, forcing the nails up into the resilient floor covering. Or if the original tiles or sheet materials were installed while there was too much moisture in the subfloor, the nails may have worked loose as the damp wood dried.

If you have an area where resilient tiles have curled at the edges or popped loose, you may have a minor plumbing leak. Moisture can also cause sheet vinyl to work loose around the perimeter of a room. Moisture in the floors of grade-level or below-grade rooms often results from poor drainage outside.

For detailed instructions on locating structural problems, and suggested remedies, see pages 87–92.

Matching replacement materials to your existing floor may be a problem, whatever the cause of damage. If you have leftover materials from the original installation, so much the better. If the flooring is relatively new, it's likely that matching materials are still available through a flooring dealer.

Even when you can find a good match, whether in your own storeroom or at a supplier, you may discover that your floor is so old or worn that new replacement material will look very conspicuous. If this is a problem, consider using replacement material in contrasting or complementary colors to create a new design—especially if you're working with individual tiles. If you can't make repairs that are visually acceptable, it may be time to consider a whole new floor covering.

A wide range of adhesives is available for making resilient floor repairs. The most common are epoxy, paste, emulsion, or some variation specified by the manufacturer of the original flooring material. For more details on the types of adhesives that are used to attach resilient flooring, ask your floor dealer.

If you're sure of the type of resilient flooring that needs repair, a flooring materials dealer should be able to tell you what type of adhesive to buy; if you're not sure, take a sample of the flooring with you. Also be prepared to tell your supplier what kind of subfloor the flooring will be attached to—concrete, felt underlayment, plywood, hardboard, or an original layer of resilient material.

Whatever adhesive you choose, be sure to check the container to find out what kind of solvent is recommended for cleaning up smudges or for removing stubborn adhesive that remains after damaged flooring has been removed. Have the solvent on hand so you can quickly remove any smudges before they dry.

Repairing minor surface damage. You can make minor surface repairs easily and effectively with tools you have on hand.

Flatten surface bubbles or blisters by cutting into them in a straight line, with a utility knife, and using a putty knife to force new adhesive under the cut, as shown below.

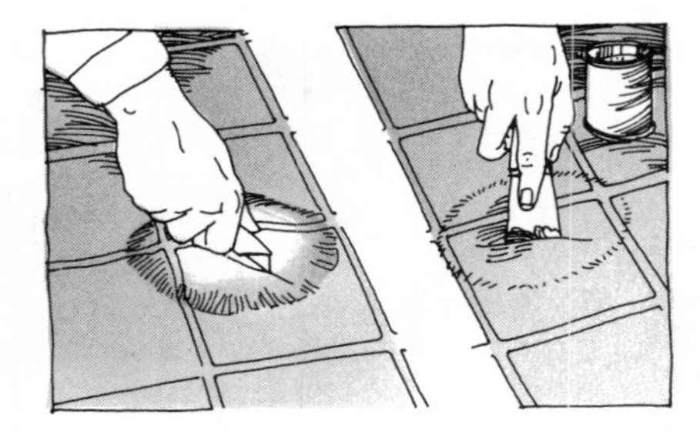

If nails have worked loose beneath a small area of the floor, reseating them may require only minor work. Try placing a block of wood over the bumps and tapping it lightly with a hammer to see if the nails can be driven flush. Be careful not to damage the surface of the floor. If the nails can't be reseated with light tapping, abandon this approach; you'll have to remove the floor covering to gain direct access to the nails and the subfloor.

Instructions on removing and replacing individual tiles or sections of sheet flooring follow.

If you're dealing with an area where tiles have curled at the edges or popped loose because of moisture below the surface, obviously you'll have to correct the moisture problem first (see page 46). Once you've done this and the floor has had time to dry thoroughly, simply scrape any loose adhesive from around the edges of the affected flooring with a knife, add new adhesive,

and press the materials firmly back into place. If you have trouble loosening a tile enough to add new adhesive, use an old iron or a propane torch to soften the old adhesive under the flooring. Warm the material, but don't heat it until it's too hot to touch. (Heat can also be used to loosen a section of damaged sheet vinyl.)

When newly fastened down, flooring material should be weighted with books, bricks, or other heavy objects until the adhesive has had time to set—usually overnight. Check the recommendations of the adhesive manufacturer for the correct setting time.

Damaged resilient tiles or whole sections of sheet vinyl can be replaced easily and neatly enough that repairs are almost invisible. Individual resilient tiles can be replaced by removing individual damaged tiles and substituting new tiles. These steps are illustrated below. Step-by-step illustrations on the next page show how to patch resilient sheet flooring.

Replacing damaged resilient tile

Soften adhesive under damaged tile using propane torch or old iron; handle torch carefully.

Pry up tile using putty knife or cold chisel. Remove old adhesive so surface of subfloor is smooth and clean.

Spread adhesive smoothly and evenly across surface of subfloor, following adhesive manufacturer's directions.

Place new tile firmly in place; use solvent to remove any adhesive smudges. Allow tile to set for recommended time.

Patching resilient sheet flooring

Cut out replacement piece large enough to cover damaged area; be sure to match any pattern.

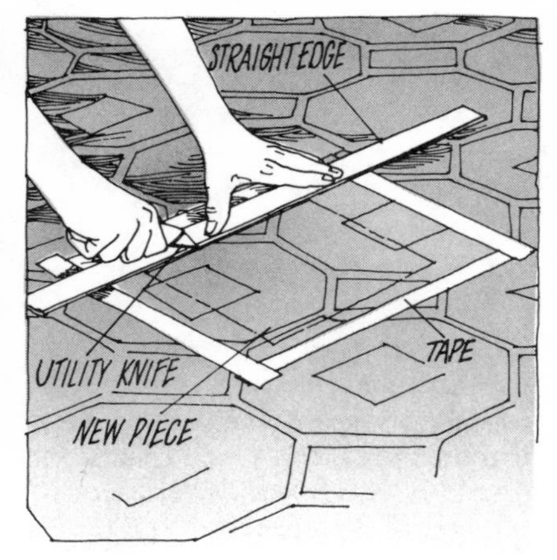

Tape new piece in place over damaged area; use utility knife and straightedge to cut through new piece and old flooring underneath.

Remove patch: use putty knife or cold chisel to pry up damaged flooring and chip out adhesive on subfloor.

Spread adhesive on back of new piece using notched trowel; then press piece firmly in place and weight it down until adhesive is set.

CARPET REPAIRS— TRICKY BUT NOT IMPOSSIBLE

Today's carpeting falls into two basic categories—conventional, which is usually laid over separate padding; and cushion-backed, which is manufactured with a rubber backing bonded to its underside and therefore needs no padding. Though both types are often made of durable synthetics, carpeting remains one of the more fragile floor coverings. It is vulnerable to snags, enduring stains, cigarette burns, and excessive wear in heavy traffic areas.

With the easy-to-use tools available today, many carpet repairs can be made by the homeowner. But before you begin any repair job, it's wise to look below the floor surface for other than obvious causes of carpet damage. A thin strip of worn carpeting might indicate loose or warped subflooring. Loose, protruding nails can cause an obvious pattern of worn "dots" in the carpeting above. For instructions on how to check the basic floor structure and for suggested remedies to structural problems, see pages 87–92.

Finding the right match

Most major carpet damage—burns, stains, or ragged tears—can be repaired by removal of the damaged section and replacement with new. The trick is to find a replacement piece that will match the surrounding carpet.

If you have leftover material from the original installation, you're ahead of the game. Still, you may be astonished to discover how much your old carpet has faded or worn when you compare it with the intended replacement.

If you don't have leftovers, check with one or more carpet suppliers to see if the particular type and color of carpet on your floor is still available. It will be helpful if you can take along even a scrap of the original carpet; otherwise, it may take considerable experimenting to find the right match.

Lacking that, consider swapping a less suitable replacement piece with a section of original carpet taken from a closet or another out-of-the-way area.

Repairing conventional carpeting

The actual task of repairing conventional wall-to-wall carpeting is not particularly difficult, but it does require careful planning, patience, and attention to detail in order to make the least conspicuous repair. Below is a list of tools you may need for the job, followed by basic instructions for patching, stitching minor tears and cuts, and trimming off and replacing pile that has minor surface damage.

Tools and supplies. For patching conventional carpeting, you'll need a knee-kicker (see illustration at right), a carpet (or utility) knife, a hammer, carpet tacks, and carpet-seam tape (available precoated or uncoated—ask your flooring dealer which is the best for your situation; if you choose uncoated, you'll also need latex seam adhesive).

To stitch up minor cuts or tears, you'll need a knee-kicker, an awl, a heavy curved needle, carpet thread that matches the color of your carpet, and latex carpet adhesive.

And to trim off minor surface damage and replace the pile, you'll need scissors with short blades, a tuft-setter, and latex carpet adhesive.

A knee-kicker can be rented from a carpet dealer (ask for directions on how to use it), and the other supplies can be purchased from the same source. A tuft-setter may have to be purchased by special order.

Patching can be tricky. Because most conventional wall-to-wall carpeting is installed with applied tension to keep it firm and taut, you'll have to relieve this tension before you can patch the area. But first cut a piece of old carpeting you have on hand into four separate strips, each strip 2 inches wide and an inch or two longer than the cuts that you plan to make around the damaged section of carpet. You'll use these strips, called "tacks," to hold the surrounding carpet in place while you repair the damaged area.

Set the knee-kicker about a foot from the area to be cut out, and nudge the knee-kicker forward until any tension in the area has been released. Place the first carpet strip upside down in front of the knee-kicker, and tack it in place (see illustration at top of next column). Using the knee-kicker in the same way on each side, tack the remaining carpet strips in a square around the damaged area.

Then, from the replacement carpet, cut out a square that is slightly larger than the damaged section (be sure to match any pattern and direction of pile), and place it over the damaged area, right side up.

If you have a wood subfloor, drive nails or long carpet tacks along one edge of the new piece, through the old carpet and into the floor. This will keep both the old and new carpet in place as you work. (If you don't have a wood subfloor, you won't be able to use nails or tacks. Instead, cut one side of the old carpet first, pry it up with your fingers, and use double-faced tape to secure the old carpet both to the floor and to the new piece; then make the remaining cuts.) Using the new piece of carpet as a guide, cut out the damaged area, taking care to cut between the rows of pile. Also be careful *not* to cut into any padding below the surface; if you do, you can use any strong household tape to mend it.

Remove the damaged carpet square and cut lengths of carpet-seam tape to place along and underneath the edges of the original carpet. If you're using uncoated tape, cover half of each strip with latex seam adhesive (read the manufacturer's directions carefully). Place the coated edge of each strip, sticky side up, under the edge of the original carpet (see illustration). Precoated tape is placed in the same way.

Apply more seam adhesive, if necessary, to the exposed portion of the tape, and press the new piece of carpeting into place. Check the drying time recommended by the adhesive manufacturer, and be sure to allow enough time for the adhesive to dry before removing the carpet tacks. When the adhesive has had time to take hold, remove the carpet tacks and brush the pile of the new piece with your fingers to blend it with the pile of the surrounding carpet.

Stitching up minor cuts or tears. Though you can use carpet-seam tape to repair minor cuts or tears in conventional wall-to-wall carpeting, a more permanent remedy is to stitch up the tear with carpet thread and reinforce it with latex seam adhesive. With either method, the procedure is the same.

Start by using the knee-kicker to reduce tension in the carpet in the corner of the room closest to the tear. Use an awl to free the carpet from the tacks in the tackless strips along the edges nearest the tear. (Tackless strips are strips of wood peppered with protruding tacks—so named because they take the place of individual tacks formerly used in carpet installation.)

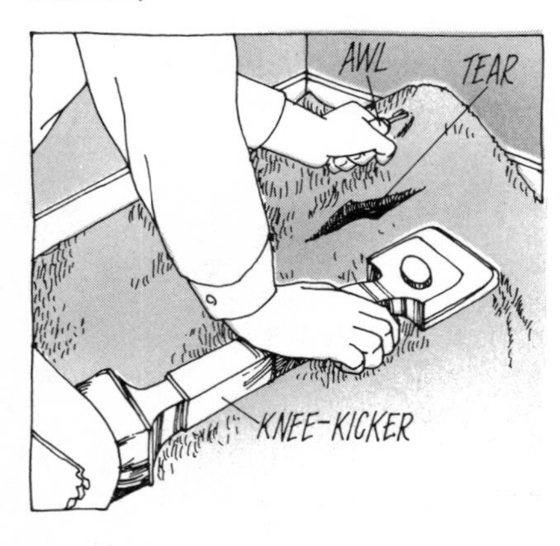

Pull back as much of the carpet as is necessary to expose the underside of the tear. Then use a heavy needle and carpet thread to sew the tear together; make stitches approximately an inch long and 1/4 inch apart. Be sure to run the needle up through the carpet backing and into the pile.

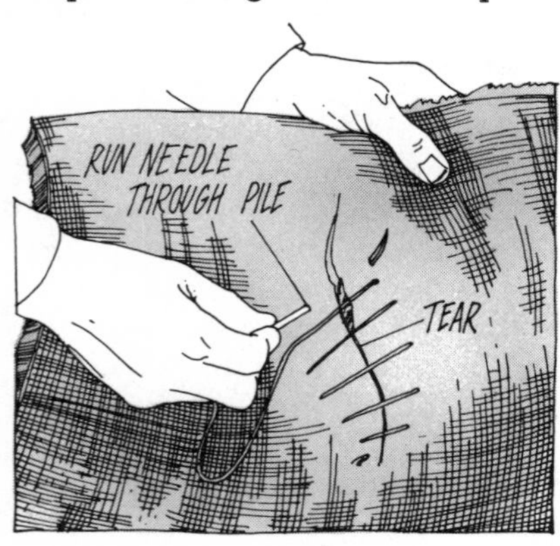

Cover the bottom of the stitched area with latex carpet adhesive. Place a piece of absorbent tissue over the adhesive and reattach the carpeting to the tackless strips with the aid of the knee-kicker.

Trimming off minor surface damage in conventional carpeting and replacing the pile can be done only if the carpet backing has not also been damaged. This repair is much more difficult than patching; it requires not only extreme patience but also great dexterity in handling tiny pieces of carpet tuft. It also requires the use of a tuft-setter.

Carefully read the directions below before deciding to tackle this kind of repair. Because different types of carpet require different approaches to replacing pile, you might want to check with your flooring dealer for further tips.

Begin by snipping off the damaged tufts down to the carpet backing. Then pull off enough individual tufts from a matching piece of carpet to fill the hole. After applying a tiny bit of seam adhesive to the exposed backing—the adhesive dries very fast, so you'll have to work quickly—fold the end of each new tuft over the tip of the tuft-setter and poke it into the carpet backing by gently tapping the handle of the tuft-setter. Place the new tufts close together and in the same direction and at the same height as the surrounding carpet pile.

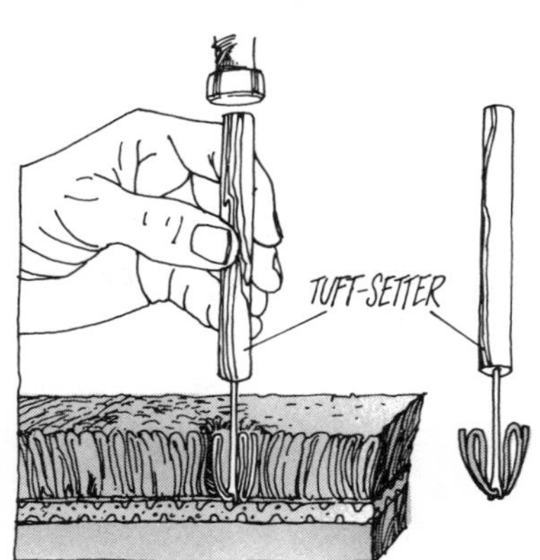

Repairing cushion-backed carpeting

Though cushion-backed carpeting was originally designed with ease of installation in mind, its characteristics also help to simplify repairs. Manufactured with a rubber backing bonded to its underside, this carpeting is laid loose or secured to the floor with double-faced tape or adhesive. Carpet laid with adhesive is more difficult to repair, because you cannot lift up the carpeting without the backing sticking to the adhesive.

Tools and supplies. Depending on the type of repair you're making (see below), you'll need a carpet or utility knife, a straightedge, double-faced tape, latex seam adhesive, a wide-bladed tool (such as a putty knife) for scraping away old adhesive, and a tuft-setter.

Patching cushion-backed carpeting is similar to patching conventional carpeting. First cut out a section of replacement carpeting slightly larger than the area of damaged carpet (match any pattern and direction of pile), and place it over the damaged area, right side up.

If the original carpet was laid loose or with double-faced tape over a wood subfloor, drive nails or long carpet tacks through one edge of the new piece and the carpet below into the subfloor; this will keep both carpet pieces in place while you work. If the carpet was laid with adhesive, the nails do not have to penetrate the subfloor. Then cut out the damaged area, using the replacement piece as a template.

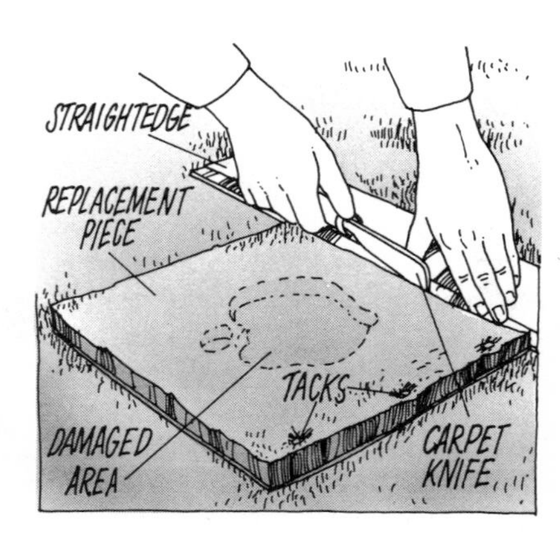

If you don't have a wood subfloor, first cut along one edge of the damaged carpet, lift it up, and use double-faced tape to secure the old carpet to the floor and to fasten the old and new carpet together. Then cut out the remaining three sides.

After removing the damaged piece of carpet, clean the subfloor, scraping away any remaining backing or adhesive.

Before removing the protective backing, slip strips of 1½ or 2-inch double-faced tape, sticky side down, halfway under each exposed edge of the old carpet.

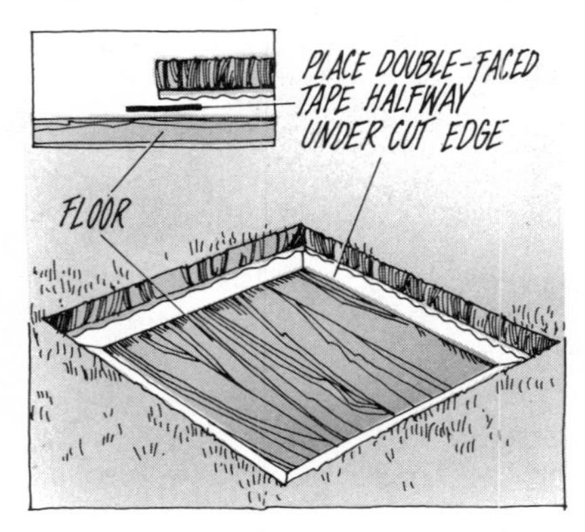

Lift up each carpet edge, remove the tape's protective covering, and press the carpet firmly to the tape. Then run a bead of latex seam adhesive along the exposed edges. Lay the replacement piece in place, pressing its edges against the tape below. Use your fingers to blend the pile of the new carpet with the old.

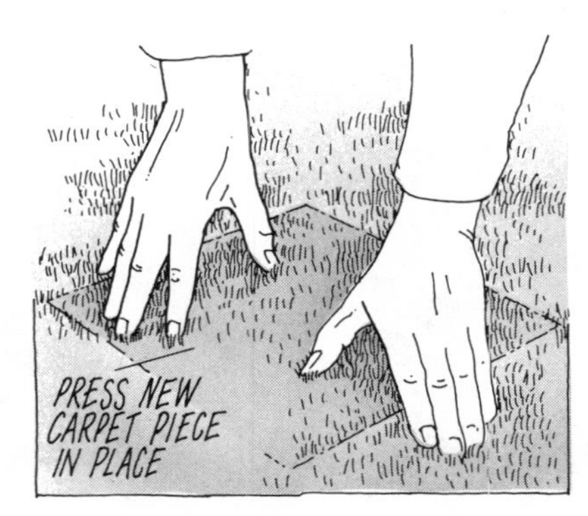

If you're using adhesive instead of tape to secure the patch, read the manufacturer's directions carefully before spreading the adhesive on the exposed subfloor.

Mending cuts or tears. As long as the carpet has not deteriorated badly around the damage, it's easy enough to mend cuts or tears in cushion-backed carpeting with tape—if the carpeting was laid loose or with double-faced tape. Because cushion-backed carpeting has only a single layer of backing (conventional carpeting has two layers), it's not strong enough to hold stitches.

Using a carpet or utility knife, slice carefully along the cut or tear to make a clean cut. Then, at opposite ends of this cut, use a straightedge to make two perpendicular cuts, forming an "H" pattern as shown. With each cut,

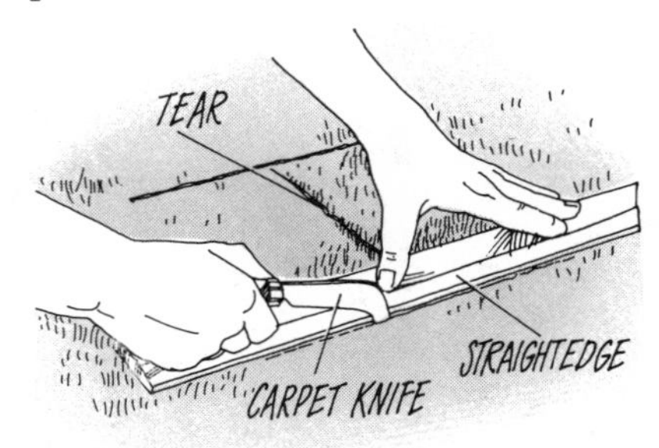

make sure you cut through the rubber backing. Fold back the cut flaps and clean the exposed subfloor.

Then cut strips of 1½ or 2-inch double-faced tape to fit the length of each perpendicular cut. Without removing its protective covering, slip the tape (sticky side down) halfway under the edge of each cut, and press the tape firmly to the floor. Lift up the outer edges of the two perpendicular cuts, remove the tape's protective covering, and press the carpet to the tape.

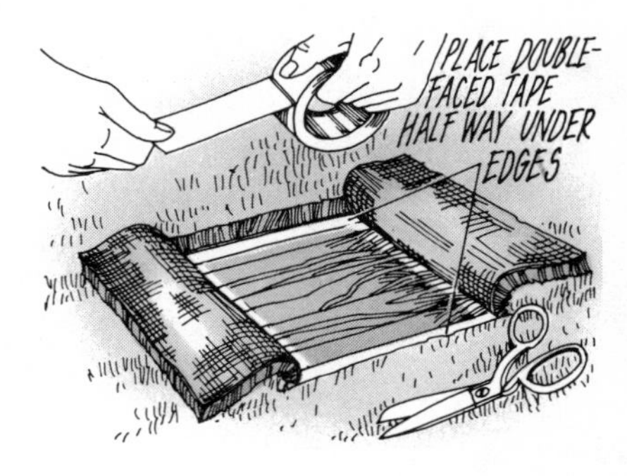

Finally, cut strips of tape to fit along the exposed ends of the folded-back carpet flaps, press each strip into place, and remove backing. Then carefully

fold each flap down onto the subfloor, pressing the edges firmly in place and blending the pile along each of the three cuts.

Trimming off minor surface damage and replacing the pile in cushion-backed carpeting is done in exactly the same way as described in the text on repairing conventional carpeting (see facing page).

INDEX

Boldface numerals refer to color photographs